1964 NASA RFP Solutions and Timing Analysis

Anatoly (Tony) Kandiew

Order this book online at www.trafford.com
or email orders@trafford.com

Most Trafford titles are also available at major online book retailers.

Print information available on the last page.

ISBN: 978-1-4907-6001-8 (sc)
ISBN: 978-1-4907-6000-1 (e)

Trafford rev. 05/20/2015

 www.trafford.com
North America & international
toll-free: 1 888 232 4444 (USA & Canada)
fax: 812 355 4082

TABLE OF CONTENTS

Prologue..vii

Introduction 1

2. 0   Benchmark Problem Solutions 2

        Problem No. 1. 2
            1A..2
            1B..7

        Problem No. 2. 11
            2A..11
            2A Timing Summary 32
            2B..33
            2B Timing Summary.. 46

        Problem No. 3. 47
            3A..47
            3A Timing Summary 54
            3B..55
            3B Timing Summary...63
            3C... 64
            3C Timing Summary 71

        Problem No. 4. 72

        Problem No. 5 76
        Problem No. 6 80
        Problem No. 7...84
            ?A.. 84
            7B.. 86

        Problem No. 8. 88
            I   Tape-to-Tape..88
            II   Tape-to-Mass Core-to-Printer............................. 90
            III   Card-to-Tape and Card...93
            IV   Remote Keyboard-Printer 94

3. 0   Definition of Terms 97

4. 0   Appendix (Representative Problem Listings) 98

        Problem No. 1A..99
        Problem No. 1B..103
        Problem No. 2A..106
        Problem No. 3A.. 111
        Problem No. 4. 113
        Problem No. 5. 117
        Problem No. ?A. 119
        Problem No. ?B..123

Epilogue..126

Prologue

Seymore Cray and Bill Norris started a data collection/engineering firm in Minneapolis after World War II called ERA (Engineering Research Associates). Seymore was the technical engineer and Bill was the salesman for that company. They were so successful that they caught the attention of a major computing company of that time: UNIVAC. First Remington Rand, then Sperry and then Univac bought them. Seymore was then working on the 1103, which became the flagship of Univac. Computers were desperately needed to compute the trajectories of guns. For all kinds of guns; on land, sea and air.

Today, as usual, the English advocate a gay pedophile who lured small boys to his flat and molested them, called Alan Turing as a lead innovator because in 1936 he conjectured a "Turing Machine" to solve certain types of algorithms. That "machine" is of dubious value, was never built, hence never worked and could not hold a candle to Seymore's inventions, who at the age of 10, using only an erector set build a machine to read paper tape and display Morse code. Turing was arrested in 1952 and while in prison killed himself in 1954! In 2013 Queen Elizabeth II, pardoned Turing posthumously. (I think it is a disgrace to the industry to promote that pedophile). Of course Seymore holds the title of "Father of the Supercomputer." Starting in 1963, when the 6600 was first announced and until today (even though he died on October 1996).

Eventually there was friction between Univac and Seymore so he and Norris left, and formed their own company called Control Data Corporation. Norris became president and Cray Senior Vice President of Development. Since Seymore worked on the Univac 1103 and Control Data moved to 501 Park Avenue, the next computer Seymore designed was the CDC 1604 (1103+501). The 1604 was his first "Supercomputer" at that time, but to cover the market needs Seymore designed also a minicomputer called the CDC 160 and 924. A short time later the 160 was upgraded to the 160A.

I joined CDC in 1961 as an instructor to teach Fortran and Assembler for both the 1604 and 160A. Since we were always short on programmers, I doubled up as a programmer and analyst. Soon I had my own territory and a few dedicated accounts: LKB (an engineering firm specializing in bridges), GASL (a consulting firm), Manhattan EYE and EAR (a Hospital) and a Hospital in the Bronx. Thus, my chief customers were located on Long Island, Manhattan and the Bronx. Very soon I became an expert on the 160A and 1604.

The story goes that one day a team of Senators visited Seymore in his Lab. Seymore showed them the 6600 and his operating system (Chippawa Operating System). But, to one of the visiting Senators, IBM whispered "parity" in his ear. So as they were leaving he raised his voice and asked "What about parity." And, Seymore replied: "Parity is for farmers." This nixed the operating system as one for "technocrats" and Control Data made the decision to develop a "user friendly" operating system called SIPROS. Thus, a team was formed in Los Angeles and charged to develop that system. After two years or about 300 man years of development there was nothing to show for, not even a scratch of paper came out of that group. Thus, CDC resorted to PLAN B.

Control Data was divided into 5 regions: East, West, Midwest, South and Washington DC. It was decided that one expert from each region should travel by car to the SIPROS group and investigate what was going on and report back to their home office. I was picked from the Eastern Region. We had to travel 300 miles each day. We received per diem and an expense account. Thus, my family and I embarked on the greatest vacation of our life. On weekends and holidays we could stay put and enjoy the area. Our first stop was Minneapolis, where we were given a one week course on the 6600. The next stop was LA.

At the designated date and time we showed up at the Sipros facility. A room was reserved for us and we were asked to be seated. Then, promptly at 9:00 o'clock the leader of the Sipros team walked in and introduced himself as Mr. Fagan. Then, he told us that we were not invited guests; therefore, as soon as he would leave, his secretary would lock us in the room. She would come back at lunch and open the door, then lock it as soon as we came back from lunch. Then, she would open the door at 5 pm for us to go home. And, so it was done!

Here we were, the smartest guys from each region neutralized by a technocrat! We began developing a plan on how to beat our confinement. First, we decided to "make friends." So, we formed a mini skills database. Mine was: Chess, Ping Pong, Cards (Bridge, Skaat, Hearts, Poker), Dancing, Stamp Collecting. We listed every skill we were good at. You never knew what could be handy in a situation like this. Then, we decided that after 5 pm we would investigate the entire building and make a map of all the offices and investigate all garbage cans for discarded listings and other papers. When we finished it was lunch time and a gorgeous lady opened the door for us. Right behind her was a tall gentleman who peeked over her shoulder and said: "Chess anyone?"

Well, here was my opportunity. I jumped up and yelled: "Here!" I walked over to the gentleman and together we went to his office. I got white and beat him. We agreed on a rematch the next day. By then lunch was over and I went back to my cave. There everybody was anxiously waiting. As soon as I walked in they asked with one voice: "Who won?" When I told them that I did, everybody was happily smiling and in good spirits. Just as the lady was about to lock us in, Fagan showed up and asked: "Who won?" With practically one voice our team yelled:"Tony won!" Fagan replied: "This is the first and last time."

When 5 pm came the lady came and let us out. We lingered for a while so that we would not arouse suspicion. After about 5 minutes the building was empty and we went to work. Amazingly we found "a ton" of listings, as we marked each location for future reference. We took the listings into our "office" and marked each stack with our name. Then we went home. The whole process took less than on our and we beat the cleaning crew by a few minutes. So, now we figured out their routine and we were sure we could repeat our excursion whenever we needed to.

Next day we pored over the listings. We separated the listings into the logical parts: monitor, utility, compiler and so on. Then, we went through the code. It was atrocious. These people did not understand the machine at all. Entire classes of instructions were missing. Then, we flowcharted each module and within a week we had a complete picture of the system. Then, we started writing scorching letters to our respective home offices.

When lunchtime came, my chess partner was waiting for me. This time I got black and lost. He used a "Kings Gambit" I had not seen before. This time, when I got back they asked who won and I told them that I had lost. Their reaction was: "Don't feel too bad, he is the Bay

Champion." But, I knew the remedy. I needed a chess book with all the King's Gambits analyzed. I told them so and they all volunteered to help me find the store and pay for the book. We had one local member from LA. After 5 pm he took me to the store and I found the book I needed. I found the opening and I liked the analysis. Next time when he would come with that opening I would beat him.

As it turned out, I won the third game and in the forth game he came back with his "King's Gambit" and I beat him. This proved to be the end of my short friendship.

Meanwhile, NASA send out an RFP (Request For Proposal) to all computer manufacturers in the USA. Companies like IBM, Burroughs, NCR, RCA, Raytheon, DEC, Univac and so on, including CDC.

The CDC RFP went to New York City, simply because the next 6600 was to be delivered to NYU. Each RFP needed to be replied within 3 months. So, the RFP arrived in New York and they did not know what to do with it. After a while, New York decided to send it to the corporate headquarters in Minneapolis. So, it arrived in Minneapolis and they did not know what to do with it either. After a while they sent it back to New York. After it sat there for a while, they remembered that I was in LA with the Sipros group. So, one day I got a frantic call from my home office and they told me about the RFP. They sent it by airplane (reserving a seat) and told me that whatever resources I needed they were mine, just get the project done on time, which was by the end of October. I received the package on October 1 and started to work on it right away.

The biggest help was the fact that we had a 6600 in the LA facility and the Fortran compiler was in good shape. So, I would solve the problem in Fortran, have it compiled and listed. Then I would optimize the generated code to fit the stack in the 6600. Then, I would use an expert typist to type my scribbles. I tried to use the format of Archimedes in solving a problem, my most favorite mathematician (he is the most plagiarized person in the world. The other two are Euler and Leibnitz). By the third week I was not only completely exhausted but my vision started to fail me, I simply could not proof read the typed result. Thus, since I had the authority, I called my home office and requested for Marcel Helou (a competent friend of mine) to come to LA and proof read the document. Marcel came on the next plane and I put him to work right away.

Within a few days the project was finished. I made 22 copies, 20 were required for NASA. One for me and one was for Marcel and as the office reference. Then, Marcel took the 20 copies to Washington, DC where they were added to the boilerplate material about the computers in question, the peripheral equipment, the management team and the history of CDC and I went to bed. (The last week was more than hectic. I had less than 6hours of sleep).

1.0 INTRODUCTION

This volume of the technical proposal concerns itself with the solution of certain segments of important problem types encountered at the Institute for Space Studies in New York City and at the Goddard Space Flight Center in Maryland. Since the arithmetic times and logical execution times of the newer generation of computing equipment are, in themselves, an inadequate measure of overall computer performance, these problem kernels are designed as a basis of relative performance analysis.

Contained herein are the calculated run times for the problem kernels on the proposed Control Data 6800, 6600, and 6400 computer systems. Since the 6600 system is operational today we have included several representative program listings with source and object coding and compilation generated on the 6600 computer in Control Data' s Los Angeles, California, facility.

Emphasis has been placed on format, equipment and problem breakdown and definition of terms to better facilitate performance analysis.

1

2.0 BENCHMARK PROBLEM SOLUTIONS

PROBLEM NO. 1A
TIROS CLOUD COVER ANALYSIS

1. Assume stored $P_{ij} = P(480, 500)$

   Logically separated into

   $P1_{i,j} = P1(480, 468)$ and

   $P2_{i,j} = P2(480, 32)$.

2. Average the sync values

   $$\overline{P2}_i = P2(480) = \frac{\sum\limits_{j=1}^{32} P2_{i,j}}{32}$$

3. Subtract out the averaged sync

   $P3_{i,k} = P3(480, 468) = P1_{i,j} - \overline{P2}_i$

4. Condense the data points

   $$P4_{k,m} = P4(240, 234) = \frac{P3_{2k-1,2m-1} + P3_{2k-1,2m} + P3_{2k,2m-1} + P3_{2k,2m}}{4}$$

5. Illustrative program:

```
      DIMENSION P(480, 500), P2B(480), P3(480, 468), P4(240, 234)
      EQUIVALENCE (P, P3, P4)
      DO 11 I = 1, 480
      P2B (I) = 0
      DO 10 J = 1, 32
10    P2B (I) = P2B (I) + P (I, J + 468)
11    P2B (I) = P2B (I)/32
      DO 15 I = 1, 480
      DO 15 J = 1, 468
15    P3B (I,J) = P (I,J) - P2B (I)
      DO 20 K = 1, 240
      DO 20 M = 1, 234
20    P4 (K, M) = (P3(2*K-1, 2*M-1) + P3(2*K-1, 2*M) + P3(2*K, 2*M-1)
     1 + P3(2*K, 2*M))/4
      END
```

<u>Method Used:</u>

1. Notation: 1 byte = 6 bits.

2. $P1_{i,j}$ = P1 (480, 47) is stored in Central Memory, 10 bytes per word.

    Note: The 47th column contains only 8 usable bytes.

    $P2_{i,j}$ = P2 (480, 4) is stored in Central Memory, 10 bytes per word.

    Note: The 4th column contains only 2 usable bytes.

3. $P2_i$ = P2 (480) = $\dfrac{\sum\limits_{j=1}^{32} P2_{i,j}}{32}$ is computed and stored in P2B.

    P2B is packed to contain 10 such bytes.

4. Subtract out the average sync: $P3_{i,j} = P1_{i,j} - P2B_i$

5. The data points are condensed in the following manner:

    $P4_{k,\ell}$ = P4 (240, 47)

    $= (P3_{2k-1, 2\ell-1} + P3_{2k-1, 2\ell} + P3_{2k, 2\ell-1} + P3_{2k, 2\ell})/4$

    $P4_{k,\ell}$ is stored in central memory.

<u>Central Memory Requirements:</u>

    P1 (480, 47) = 22, 560 CM Locations
    P2 (480, 4) = 1, 920 CM Locations
    P2B (480)   =   480 CM Locations

        TOTAL = 24, 960 CM Locations

6800 Timing for Problem 1A

1. Notation:

$T_M = 25 * 10^{-9}$ sec	T_{CP} = Central processor time
$T_{MJ} = 250 * 10^{-9}$ sec	I = Number of instructions executed
$P = 100$	
	T = Total job time
$T_{MT} = \begin{cases} 100 * T_{MJ} \text{ average} \\ 200 * T_{MJ} \text{ maximum} \end{cases}$	

2. Formulas:

$$T_{CP} = T_M \left[480 \left(77 + (12 \cdot 28 + 16) + \frac{1}{2}(13 + 47(28)) \right) + 118 \right]$$

$$I = \left[480 \left(42 + (12 \cdot 14 + 5) + \frac{1}{2}(4 + 47(16)) \right) + 31 \right]$$

$$T = T_{CP} + T_{MT}$$

3. Results

T_{CP} (ms)	I	T (ms)
13.125	284,671	13.149

4

6600 Timing for Problem 1A

1. Notation

$T_M = 100 * 10^{-9}$ sec	T_{CP} = Central processor time
$T_{MJ} = 1 * 10^{-6}$ sec	I = Number of instructions executed
$P = 100$	
$T_{MT} = \begin{cases} 100 \cdot T_{MJ} \text{ average} \\ 200 \cdot T_{MJ} \text{ maximum} \end{cases}$	T = Total job time

2. Formulas

$$T_{CP} = T_M \left[480 \left(77 + (12 \cdot 28 + 16) + \frac{1}{2}\left(13 + 47(28) \right) \right) + 118 \right]$$

$$I = \left[480 \left(42 + (12 \cdot 14 + 5) + \frac{1}{2}\left(4 + 47(16) \right) \right) + 31 \right]$$

$$T = T_{CP} + T_{MT}$$

3. Results

T_{CP} (ms)	I	T (ms)
52. 488	284, 671	52. 588

5

1. Notation

$T_M = 100 * 10^{-9}$ sec $T_{MJ} = 1 * 10^{-6}$ sec $P = 100$ $T_{MT} = \begin{cases} 100 \cdot T_{MJ} \text{ average} \\ 200 \cdot T_{MJ} \text{ maximum} \end{cases}$	T_{CP} = Central processor time I = Number of instructions executed T = Total job time

2. Formulas

$$T_{CP} = T_M \left[480 \left(214 + (12 \cdot 73 + 25) + \frac{1}{2}\left(27 + 47(86) \right) \right) + 156 \right]$$

$$I = \left[480 \left(42 + (12 \cdot 14 + 5) + \frac{1}{2}\left(4 + 47(16) \right) \right) + 31 \right]$$

$$T = T_{CP} + T_{MT}$$

3. Results

T_{CP} (ms)	I	T (ms)
151. 191	284, 671	151. 291

6

1. Assume stored the reduced picture $R_{i,j}$ = R (240, 240) where the character packing is the same as in Problem 1A.

2. Select out an enclosed subset of the picture R..
    $R1_{k,p}$ = R1 (200, 200) where the subscripts k, p cover the following ranges of i, j respectively.

    k = 21, 200 of i

    p = 21, 220 of j

3. From R1 (200, 200) pick out the 100 square element subset, $R2_{m,n}$ = R2 (20, 20), 20 characters on a side. M and n will both range from 1 to 10. For a given m, n the k, p values will have the range

    k = 20 * m + 1 ,   20 * m + 20

    p = 20 * n + 1 ,   20 * n + 20

4. For each $R2_{m,n}$ count the number of grey level values, occurring for each value 0, 1, 2, $\cdots$ 31 or 32 through 63 and store this count in S (m, n, k) where k ranges from 1 to 33.

5. Illustrative program

```
        DIMENSION GREY (32), R(240, 240), S(10, 10, 33), R1(200, 200)
        EQUIVALENCE (R (21, 21), R1(1, 1))
        DO 10 I = 1, 32
  10    GREY (I) = I - 1
        DO 100 M = 1, 10
        DO 100 N = 1, 10
        DO 30 K = 20 * M + 1, 20 * M + 20
        DO 30 L = 20 * N + 1, 20 * N + 20
        DO 20 J = 1, 32
        IF (R (K, L) - GREY (J)) 20, 21, 20
  20    CONTINUE
        S (M, N, 33) = S (M, N, 33) + 1
        GO TO 30
  21    S (M, N, J) = S (M, N, J) + 1
  30    CONTINUE
 100    CONTINUE
```

<u>6800 Timing for Problem 1B</u>

1. <u>Notation</u>

$T_M = 25 * 10^{-9}$ sec $\quad$ $T_{MJ} = 250 * 10^{-9}$ sec $\quad$ $P = 100$ $\quad$ $T_{MT} = \begin{cases} 100 * T_{MJ} \text{ average} \\ 200 * T_{MJ} \text{ maximum} \end{cases}$	T_{CP} = Central processor time $\quad$ I = Number of instructions $\quad$ executed $\quad$ T = Total job time

2. <u>Formulas</u>

$$T_{CP} = T_M \left[119 + 8(24) + 32 \left(13 + 100 \left(14 + 40(135) \right) \right) + 100 \left(12 + 32(12) \right) \right]$$

$$I = \left[36 + 8(10) + 32 \left(6 + 100 \left(5 + 40(48) \right) \right) + 100 \left(6 + 32(4) \right) \right]$$

$$T = T_{CP} + T_{MT}$$

3. <u>Results</u>

T_{CP} (ms)	I	T (ms)
434. 128	6, 173, 708	434. 153

8

6600 Timing for Problem 1B

1. Notation

$T_M = 100 * 10^{-9}$ $T_{MJ} = 1 * 10^{-6}$ sec $P = 100$ $T_{MT} = \begin{cases} 100 \cdot T_{MJ} \text{ average} \\ 200 \cdot T_{MJ} \text{ maximum} \end{cases}$	T_{CP} = Central processor time I = Number of instructions executed T = Total job time

2. Formulas

$$T_{CP} = T_M \left[119 + 8(24) + 32\left(13 + 100\left(14+40(135)\right)\right) + 100\left(12+32(12)\right) \right]$$

$$I = \left[36 + 8(10) + 32\left(6 + 100\left(5+40(48)\right)\right) + 100\left(6+32(4)\right) \right]$$

$$T = T_{CP} + T_{MT}$$

3. Results

T_{CP} (ms)	I	T (ms)
1736.513	6,173,708	1736.613

6400 Timing for Problem 1B

1. Notation

$$T_M = 100 * 10^{-9} \text{ sec}$$

$$T_{MJ} = 1 * 10^{-6} \text{ sec}$$

$$P = 100$$

$$T_{MT} = \begin{cases} 100 \cdot T_{MJ} \text{ average} \\ 200 \cdot T_{MJ} \text{ maximum} \end{cases}$$

T_{CP} = Central processor time

I = Number of instructions executed

T = Total job time

2. Formulas

$$T_{CP} = T_M \left[185 + 8\,(61) + 32 \left(36 + 100 \left(32 + 40(277) \right) \right) + 100 \left(28 + 32(22) \right) \right]$$

$$I = \left[36 + 8\,(10) + 32 \left(6 + 100 \left(5 + 40(48) \right) \right) + 100 \left(6 + 32(4) \right) \right]$$

$$T = T_{CP} + T_{MT}$$

3. Results

T_{CP} (ms)	I	T (ms)
3563. 343	6, 173, 708	3563. 443

PROBLEM NO. 2A
SOLUTION OF ELLIPTIC PARTIAL DIFFERENTIAL EQUATIONS

1. Obtain the time required to perform $P = 100$ iterations of the following:

$$U_{i,j,k}^{P} = W_{i,j,k}^{1} \cdot U_{i+1,j,k}^{(P-1)} + W_{i,j,k}^{2} \cdot U_{i-1,j,k}^{(P)} + W_{i,j,k}^{3} \cdot U_{i,j+1,k}^{(P-1)}$$

$$+ W_{i,j,k}^{4} \cdot U_{i,j-1,k}^{(P)} + W_{i,j,k}^{5} \cdot U_{i,j,k+1}^{(P-1)} + W_{i,j,k}^{6} \cdot U_{i,j,k-1}^{(P)}$$

$$+ W_{i,j,k}^{7} \cdot U_{i,j,k}^{(P-1)} + W_{i,j,k}^{8} \cdot f_{i,j,k}$$

for: $\quad 2 \le i \le n_x - 1 , \quad 2 \le j \le n_y - 1 , \quad 2 \le k \le n_z - 1$

$$d^{P} = \frac{1}{n_x \, n_y \, n_z} \sum_{k=2}^{n_z-1} \sum_{j=2}^{n_y-1} \sum_{i=2}^{n_x-1} \left| U_{i,j,k}^{(P)} - U_{i,j,k}^{(P-1)} \right|$$

2. Illustrative Problem:

```
          DO 1 ITER = 1, 100
          D = 0.0
          DO 1 K = 2, NZ-1
          DO 1 J = 2, NY-1
          DO 1 I = 2, NX-1
          TEMP  = W1(I, J, K) * U(I+1, J, K) + W2(I, J, K) * U(I-1, J, K)
                  + W3(I, J, K) * U(I, J+1, K) + W4(I, J, K) * U(I, J-1, K)
                  + W5(I, J, K) * U(I, J, K+1) + W6(I, J, K) * U(I, J, K-1)
                  + W7(I, J, K) * U(I, J, K) + W8(I, J, K) * F(I, J, K)
          D = D + ABSF ((TEMP - U(I, J, K))
     1    U(I, J, K) = TEMP
     2    D = D/FLOATF (NX * NY * NZ)
```

1. Case No. 1:

    All data is kept in core storage. ($N_x = 25$, $N_y = 25$, $N_z = 5$)

2. Notation:

$T_M = 25 * 10^{-9}$ sec $T_{MJ} = 250 * 10^{-9}$ sec $P = 100$ $T_{MT} = \begin{cases} 100 * T_{MJ} \text{ average} \\ 200 * T_{MJ} \text{ maximum} \end{cases}$	T_{CP} = Central processor time I = Number of instructions executed T = Total job time

3. Formulas:

$$T_{CP} = T_M * P * \left[75 + (N_z-2)\left(138 + (N_y-2)\left(50 + (N_x-2)(91)\right)\right)\right]$$
$$I = P * \left[20 + (N_z-2)\left(32 + (N_y-2)\left(16 + (N_x-2)(48)\right)\right)\right]$$
$$T = T_{CP} + T_{MT}$$

4. Results:

N_x	N_y	N_z	T_{CP} (sec)	I	T (sec)
25	25	5	0.371	773×10^4	0.371

1. Case No. 2:

   Data is kept in core and extended memory. (N_x = 25, N_y = 25, N_z = 25)

   a. In Core:

      U(I, J, K)
      W1(I, J, 1) through W1(I, J, 21)
      W2(I, J, 1) through W2(I, J, 21)
      W3(I, J, 1) through W3(I, J, 21)
      W4(I, J, 1) through W4(I, J, 21)
      W5(I, J, 1) through W5(I, J, 21)
      W6(I, J, 1) through W6(I, J, 21)
      W7(I, J, 1) through W7(I, J, 21)
      W8(I, U, 1) through W8(I, J, 21)
      F(I, J, 1)　 through F(I, J, 21)

   b. In Extended Core:

      The remaining planes of W and F and the first four planes of W through F.

   c. Two passes are required to bring in the necessary data per iteration:

      Pass No. 1: Read planes 22 through 25 of W and F into planes 1 through 4 of W and F.
      Pass No. 2: Restore planes 1 through 4 of W and F.

2. Notation:

$T_M = 25 * 10^{-9}$ sec	T_{CP} = Central processor time
$T_{MJ} = 250 * 10^{-9}$ sec	I = Number of instructions executed
P = 100	
$T_{MT} = \begin{cases} 100 * T_{MJ} \text{ average} \\ 200 * T_{MJ} \text{ maximum} \end{cases}$	T_{ECT} = Extended core transfer time with access
$T_{EC} = (0.8 + 1.6) * 10^{-6}$ sec	T = Total job time

3. Formulas:

$$T_{CP} = T_M * P * \left[75 + (N_z - 2)\left(138 + (N_y - 2)\left(50 + (N_x - 2)(91)\right)\right)\right]$$

$$I = P * \left[20 + (N_z - 2)\left(32 + (N_y - 2)\left(16 + (N_x - 2)(48)\right)\right)\right]$$

13

$$T_{ECT} = 2 * P * \left[T_{EC} + T_M * N_x * N_y * 4 \right]$$

$$T = T_{CP} + T_{ECT} + T_{MT}$$

4. Results:

N_x	N_y	N_z	T_{CP} (sec)	I	T_{ECT} (sec)	T (sec)
25	25	25	2.842	593×10^5	0.013	2.855

14

1. Case No. 3:

    Data is kept in core, extended core and disk. ($N_x = 50$, $N_y = 50$, $N_z = 50$)

    a. In Core:

    All of core is used as work area to contain
    7 planes of U(I, J) and
    5 planes of W1(I, J) through F(I, J)

    b. In Extended Core:

    All of U(I, J, K) is kept permanently in extended core. Also 38 planes
    of W1 through W8 and F are kept in extended core, e.g.:

       W1(I, J, 1) through W1(I, J, 38)
       W2(I, J, 1) through W2(I, J, 38)
       W3(I, J, 1) through W3(I, J, 38)
       W4(I, J, 1) through W4(I, J, 38)
       W5(I, J, 1) through W5(I, J, 38)
       W6(I, J, 1) through W6(I, J, 38)
       W7(I, J, 1) through W7(I, J, 38)
       W8(I, J, 1) through W8(I, J, 38)
       F(I, J, 1) through F(I, J, 38)

    c. On the Disk:

    The remaining 12 planes of W1 through W8 and F are kept on
    2 disks.
    40 planes of W1(I, J) through F(I, J) can be kept on 2 disks on
    1 position.
    Therefore 4 planes of W1 through F can be kept on 1 position.
    Hence 3 positions on the disk are required to hold the outstanding
    data.

2. Notation:

$$T_M = 25 * 10^{-9} \text{ sec}$$

$$T_{MJ} = 250 * 10^{-9} \text{ sec}$$

$$P = 100$$

$$T_{MT} = \begin{cases} 100 * T_{MJ} \text{ average} \\ 200 * T_{MJ} \text{ maximum} \end{cases}$$

$$T_{EC} = (0.8 + 1.6) * 10^{-6} \text{ sec}$$

$$T_{DL} = 33.5 \text{ ms}$$

$$T_D = 7.91 \text{ } \mu s$$

$$T_{DP} = (3 * 67 + T_{DL}) \text{ ms}$$

$$P_O = 6$$

$$W_O = 385 * T_M$$

$$I_O = \left[385 + 108(P_O - 1)\right] * T_M$$

T_{CP} = Central processor time

I = Number of instructions executed

T_{ECT} = Extended core transfer time with access

T_{DT} = Disk transfer time

T_{ITRK} = Time to compute k planes

K = 4 for disk

$T_{I/O}$ = Time to set up disk I/O request

T_{OV} = Overlap time of I/O and compute

3. Formulas:

$$T_{CP} = T_M * P * \left[75 + (N_z - 2)\left(138 + (N_y - 2)\left(50 + (N_x - 2)\,(91)\right)\right)\right]$$

$$I = P * \left[20 + (N_z - 2)\left(32 + (N_y - 2)\left(16 + (N_x - 2)\,(48)\right)\right)\right]$$

$$T_{ECT} = P * \left[T_{EC} + T_M * N_x * N_y * 5 * 10\right]\left(\frac{N_z - 5}{5}\right)$$

$$T_{DT} = P * \left[\frac{1}{2} T_{DP} + T_{DL} + \frac{1}{2} T_D * N_x \cdot N_y * 4 * 9\right] * 3$$

$$T_{ITRK} = K * \left[138 + (N_y - 2)\left(50 + (N_x - 2)\,(91)\right)\right] * T_M$$

$$T_{I/O} = \left[2 * \left(I_O + T_{MT} + W_O\right)\right] * 3 * P$$

$$T_{OV} = T_{ITRK} \quad \text{where } 0 \leq T_{OV} \leq T_{DP}$$

$$T = T_{CP} + T_{ECT} + \left(T_{DT} - P * \left(\frac{N_z}{K}\right) * T_{ITRK}\right) + T_{I/O} + T_{MT}$$

4. Results:

N_x	N_y	N_z	T_{CP} (sec)	I	T_{ECT} (sec)	T_{DT} (sec)	$T_{I/O}$ (sec)	T_{ITRK} (sec)	T (sec)
50	50	50	25.464	534×10^6	2.814	152.010	0.035	0.021	154.073

1. Case No. 4:

    Data is kept in core, extended core and disk. (N_x = 100, N_y = 100, N_z = 100)

    a. In Core:

        All of core is used as work area to contain:
        3 planes of U(I, J)
        1 plane each of W1 through F.

    b. In Extended Core:

        U(I, J, K) is kept permanently in extended core.

    c. On Disks:

        All W1 through W8 and F are kept on 2 disks.
        1 plane of W1 (I, J) through F(I, J) can be kept on 2 disks on 1 position.
        Therefore 100 positions on both disks are required to hold required data.

2. Notation:

$T_M = 25 * 10^{-9}$ sec	$I_O = \left[385 + 108 \ (P_O - 1)\right] * T_M$
$T_{MJ} = 250 * 10^{-9}$ sec	$K = 1$
$T_{MT} = \begin{cases} 100 * T_{MJ} \text{ average} \\ 200 * T_{MJ} \text{ maximum} \end{cases}$	T_{CP} = Central processor time
$T_{EC} = (0.8 + 1.6) * 10^{-6}$ sec	I = Number of instructions executed
$T_{DL} = 33.5$ ms	T_{ECT} = Extended core transfer time with access
$T_D = 7.91 \ \mu s$	T_{DT} = Disk transfer time
$T_{DP} = \left(3 * 67 + T_{DL}\right)$ ms	T_{ITRK} = Time to compute 1 plane
$P_O = 6$	$T_{I/O}$ = Time to set up disk I/O request
$W_O = 385 * T_M$	
$P = 100$	T_{OV} = Overlap time of I/O and compute

18

3. Formulas:

$$T_{CP} = T_M * P * \left[75 + (N_z-2)\left(138 + (N_y-2)\left(50 + (N_x-2)(91)\right)\right)\right]$$

$$I = P * \left[20 + (N_z-2)\left(32 + (N_y-2)\left(16 + (N_x-2)(48)\right)\right)\right]$$

$$T_{ECT} = P * 2 * \left[T_{EC} + T_M * N_x * N_y * 2\right] * (N_z)$$

$$T_{DT} = P * \left[T_{DP} + T_{DL} + \frac{1}{2} * T_D * N_x * N_y * 9\right] * (N_z)$$

$$T_{ITRK} = \left[138 + (N_y-2)\left(50 + (N_x-2)(91)\right)\right] * T_M$$

$$T_{I/O} = P * \left[2 * \left(I_O + T_{MT} + W_O\right) * N_z\right]$$

$$T_{OV} = T_{ITRK} \quad \text{where } 0 \le T_{OV} \le T_{DP}$$

$$T = T_{CP} + T_{ECT} + \left(T_{DT} - P * N_z * T_{ITRK}\right) + T_{I/O} + T_{MT}$$

4. Results:

N_x	N_y	N_z	T_{CP} (sec)	I	T_{ECT} (sec)	T_{DT} (sec)	$T_{I/O}$ (sec)	T_{ITRK} (sec)	T (sec)
100	100	100	215.355	453×10^7	10.048	6,239.500	1.209	.022	6,245.612

6600 Timing for Problem No. 2A

1. Case No. 1: ($N_x = 25$, $N_y = 25$, $N_z = 5$)

2. Notation:

$T_M = 100 * 10^{-9}$ sec	T_{CP} = Central processor time
$T_{MJ} = 1 * 10^{-6}$ sec	I = Number of instructions executed
P = 100	
$T_{MT} = \begin{cases} 100 \cdot T_{MJ} \text{ average} \\ 200 \cdot T_{MJ} \text{ maximum} \end{cases}$	T = Total job time

3. Formulas:

$$T_{CP} = T_M * P * \left[75 + (N_z-2)\left(138 + (N_y-2)\left(50 + (N_x-2)(91)\right)\right)\right]$$

$$I = P * \left[20 + (N_z-2)\left(32 + (N_y-2)\left(16 + (N_x-2)(48)\right)\right)\right]$$

$$T = T_{CP} + T_{MT}$$

4. Results:

N_x	N_y	N_z	T_{CP} (sec)	I	T (sec)
25	25	5	1.484	773×10^4	1.484

20

6600 Timing for Problem No. 2A

1. Case No. 2: $(N_x = 25, N_y = 25, N_z = 25)$

2. Notation:

$T_M = 100 * 10^{-9}$ sec	T_{CP} = Control processor time
$T_{MJ} = 1 * 10^{-6}$ sec	I = Number of instructions executed
P = 100	
$T_{MT} = \begin{cases} 100 * T_{MJ} \text{ average} \\ 200 * T_{MJ} \text{ maximum} \end{cases}$	T_{ECT} = Extended core transfer time with access
$T_{EC} = (1.6 + 3.2) * 10^{-6}$ sec	T = Total job time

3. Formulas:

$$T_{CP} = T_M * P * \left[75 + (N_z - 2) \left(138 + (N_y - 2) \left(50 + (N_x - 2)(91) \right) \right) \right]$$

$$I = P * \left[20 + (N_z - 2) \left(32 + (N_y - 2) \left(16 + (N_x - 2)(48) \right) \right) \right]$$

$$T_{ECT} = 2 * P * \left[T_{EC} + T_M * N_x * N_y * 4 \right]$$

$$T = T_{CP} + T_{ECT} + T_{MT}$$

4. Results:

N_x	N_y	N_z	T_{CP} (sec)	I	T_{ECT} (sec)	T (sec)
25	25	25	11.369	593×10^5	.051	11.420

6600 Timing for Problem No. 2A

1. Case No. 3: ($N_x = 50$, $N_y = 50$, $N_z = 50$)

2. Notation:

$$T_M = 100 * 10^{-9} \text{ sec}$$

$$T_{MJ} = 1 * 10^{-6} \text{ sec}$$

$$P = 100$$

$$T_{MT} = \begin{cases} 100 * T_{MJ} \text{ average} \\ 200 * T_{MJ} \text{ maximum} \end{cases}$$

$$T_{EC} = (1.6 + 3.2) * 10^{-6} \text{ sec}$$

$$T_{DL} = 33.5 \text{ ms}$$

$$T_D = 7.91 \text{ μs}$$

$$T_{DP} = (3 * 67 + T_{DL}) \text{ms}$$

$$W_O = 385 * T_M$$

$$I_O = \left[385 + 108 \, (P_O - 1)\right] * T_M$$

T_{CP} = Central processor time

I = Number of instructions executed

T_{ECT} = Extended core transfer time with access

T_{DT} = Disk transfer time

T_{ITRK} = Time to compute k planes

k = 4 for disk

$T_{I/O}$ = Time to set up disk I/O request

T_{OV} = Overlap time of I/O and compute

$P_O = 6$

3. Formulas:

$$T_{CP} = T_M * P * \left[75 + (N_z - 2)\left(138 + (N_y - 2)\left(50 + (N_x - 2)\,(91)\right)\right)\right]$$

$$I = P * \left[20 + (N_z - 2)\left(32 + (N_y - 2)\left(16 + (N_x - 2)\,(91)\right)\right)\right]$$

$$T_{ECT} = P * \left[T_{EC} + T_M * N_x * N_y * 5 * 10\right]\left(\frac{N_z - 5}{5}\right)$$

$$T_{DT} = P * \left[\tfrac{1}{2} T_{DP} + T_{DL} + \tfrac{1}{2} T_D * N_x * N_y * 4 * 9\right] * 3$$

$$T_{ITRK} = K * \left[138 + (N_y - 2)\left(50 + (N_x - 2)\,(91)\right)\right] * T_M$$

$$T_{I/O} = P * \left[2 * \left(I_O + T_{MT} + W_O\right)\right] * 3$$

$$T_{OV} = T_{ITRK}, \quad \text{where } 0 \leq T_{OV} \leq T_{DP}$$

$$T = T_{CP} + T_{ECT} + \left(T_{DT} - P * \left(\frac{N_z}{K}\right) * T_{ITRK}\right) + T_{I/O} + T_{MT}$$

4. Results:

N_x	N_y	N_z	T_{CP} (sec)	I	T_{ECT} (sec)	T_{DT} (sec)	$T_{I/O}$ (sec)	T_{ITRK} (sec)	T (sec)
50	50	50	101.857	534×10^6	11.254	152.010	.145	.084	160.266

6600 Timing for Problem No. 2A

1. Case No. 4: ($N_x = 100$, $N_y = 100$, $N_z = 100$)

2. Notation:

$$T_M = 100 * 10^{-9} \text{ sec}$$

$$T_{MJ} = 1 * 10^{-6} \text{ sec}$$

$$T_{MT} = \begin{cases} 100 * T_{MJ} \text{ average} \\ 200 * T_{MJ} \text{ maximum} \end{cases}$$

$$T_{EC} = (1.6 + 3.2) * 10^{-6} \text{ sec}$$

$$T_{DL} = 33.5 \text{ ms}$$

$$T_D = 7.91 \text{ μs}$$

$$T_{DP} = (3 * 67 + T_{DL}) \text{ ms}$$

$$P_O = 6$$

$$W_O = 385 * T_M$$

$$P = 100$$

$$I_O = \left[385 + 108\,(P_O - 1)\right] * T_M$$

$$K = 1$$

T_{CP} = Control processor time

I = Number of instructions executed

T_{ECT} = Extended core transfer time with access

T_{DT} = Disk transfer time

T_{ITRK} = Time to compute 1 plane

$T_{I/O}$ = Time to set up disk I/O request

T_{OV} = Overlap time of I/O and compute

3. Formulas:

$$T_{CP} = T_M * P * \left[75 + (N_z - 2)\left(138 + (N_y - 2)\left(50 + (N_x - 2)\,(91)\right)\right)\right]$$

$$I = P * \left[20 + (N_z - 2)\left(32 + (N_y - 2)\left(16 + (N_x - 2)\,(48)\right)\right)\right]$$

$$T_{ECT} = P * 2 * \left[T_{EC} + T_M * N_x * N_y * 2\right] * (N_z)$$

$$T_{DT} = P * \left[T_{DP} + T_{DL} + \tfrac{1}{2} T_D * N_x * N_y * 9\right] * (N_z)$$

$$T_{ITRK} = \left[138 + (N_y - 2)\left(50 + (N_x - 2)\,(91)\right)\right] * T_M$$

$$T_{I/O} = P * \left[2 * (I_O + T_{MT} + W_O) * N_z\right]$$

$$T_{OV} = T_{ITRK} \quad \text{where } 0 \leq T_{OV} \leq T_{DP}$$

$$T = T_{CP} + T_{ECT} + \left(T_{DT} - P * N_z * T_{ITRK}\right) + T_{I/O} + T_{MT}$$

4. Results:

N_x	N_y	N_z	T_{CP} (sec)	I	T_{ECT} (sec)	T_{DT} (sec)	$T_{I/O}$ (sec)	T_{ITRK} (sec)	T (sec)
100	100	100	861.422	453×10^7	40.096	6239.500	4.836	.088	6265.854

6400 Timing for Problem No. 2A

1. <u>Case No. 1</u>: $(N_x = 25, N_y = 25, N_z = 5)$

2. Notation:

$T_M = 100 * 10^{-9}$ sec $T_{MJ} = 1 * 10^{-6}$ sec $P = 100$ $T_{MT} = \begin{cases} 100 * T_{MJ} \text{ average} \\ 200 * T_{MJ} \text{ maximum} \end{cases}$	T_{CP} = Central processor time I = Number of instructions executed T = Total job time

3. Formulas:

$$T_{CP} = T_M * P * \left[104 + (N_z-2)\left(334 + (N_y-2)\left(84 + (N_x-2)(715)\right)\right)\right]$$
$$I = P * \left[20 + (N_z-2)\left(32 + (N_y-2)\left(16 + (N_x-2)(48)\right)\right)\right]$$
$$T = T_{CP} + T_{MT}$$

4. Results:

N_x	N_y	N_z	T_{CP} (sec)	I	T (sec)
25	25	5	11.416	773×10^4	11.416

6400 Timing for Problem No. 2A

1. Case No. 2: ($N_x = 25$, $N_y = 25$, $N_z = 25$)

2. Notation:

$T_M = 100 * 10^{-9}$ sec $T_{MJ} = 1 * 10^{-6}$ sec $P = 100$ $T_{MT} = \begin{cases} 100 * T_{MJ} \text{ average} \\ 200 * T_{MJ} \text{ maximum} \end{cases}$ $T_{EC} = (1.6 + 3.2) * 10^{-6}$ sec	T_{CP} = Central processor time I = Number of instructions executed T_{ECT} = Extended core transfer time with access T = Total job time

3. Formulas:

$$T_{CP} = T_M * P * \left[104 + (N_z-2)\left(334 + (N_y-2)\left(84 + (N_x-2)(715)\right)\right)\right]$$
$$I = P* \left[20 + (N_z-2)\left(32 + (N_y-2)\left(16 + (N_x-2)(48)\right)\right)\right]$$
$$T_{ECT} = 2 * P * \left[T_{EC} + T_M * N_x * N_y * 4\right]$$
$$T = T_{CP} + T_{ECT} + T_{MT}$$

4. Results:

N_x	N_y	N_z	T_{CP} (sec)	I	T_{ECT} (sec)	T (sec)
25	25	25	87.516	593×10^5	.051	87.567

6400 Timing for Problem No. 2A

1. Case No. 3: ($N_x = 50$, $N_y = 50$, $N_z = 50$)

2. Notation:

$T_M = 100 * 10^{-9}$ sec	T_{CP} = Central processor time
$T_{MJ} = 1 * 10^{-6}$ sec	I = Number of instructions executed
$P = 100$	
$T_{MT} = \begin{cases} 100 * T_{MJ} \text{ average} \\ 200 * T_{MJ} \text{ maximum} \end{cases}$	T_{ECT} = Extended core transfer time with access
$T_{EC} = (1.6 + 3.2) * 10^{-6}$ sec	T_{DT} = Disk transfer time
$T_{DL} = 33.5$ ms	T_{ITRK} = Time to compute k planes
$T_D = 7.91$ μs	$T_{I/O}$ = Time to set up disk I/O request
$T_{DP} = (3 * 67 + T_{DL})$ ms	T_{OV} = Overlap time of I/O and compute
$W_O = 495 * T_M$	$k = 4$
$I_O = \left[495 + 184\,(P_O-1)\right] * T_M$	$P_O = 6$

3. Formulas:

$$T_{CP} = T_M * P * \left[104 + (N_z-2)\left(334 + (N_y-2)\left(84 + (N_x-2)\,(715)\right)\right)\right]$$

$$I = P * \left[20 + (N_z-2)\left(32 + (N_y-2)\left(16 + (N_x-2)\,(48)\right)\right)\right]$$

$$T_{ECT} = P * \left[T_{EC} + T_M * N_x + N_y * 5 * 10\right]\left(\frac{N_z-5}{5}\right)$$

$$T_{DT} = P * \left[\frac{1}{2}T_{DP} + T_{DL} + \frac{1}{2}T_D * N_x * N_y * 4 * 9\right] * 3$$

$$T_{ITRK} = K * \left[334 + (N_y-2)\left(84 + (N_x-2)\,(715)\right)\right] * T_M$$

$$T_{I/O} = P * \left[2 * \left(I_O + T_{MT} + W_O\right)\right] * 3$$

$$T_{OV} = T_{ITRK}, \quad \text{where } 0 \le T_{OV} \le T_{DP}$$

$$T = T_{CP} + T_{ECT} + \left(T_{DT} - P * \left(\frac{N_z}{K}\right) * T_{ITRK}\right) + T_{I/O} + T_{MT}$$

4. Results:

N_x	N_y	N_z	T_{CP} (sec)	I	T_{ECT} (sec)	T_{DT} (sec)	$T_{I/O}$ (sec)	T_{ITRK} (sec)	T (sec)
50	50	50	792.829	534×10^6	11.254	152.010	.179	.661	812.709

6400 Timing for Problem No. 2A

1. <u>Case No. 4</u>: ($N_x = 100$, $N_y = 100$, $N_z = 100$)

2. Notation:

$T_M = 100 * 10^{-9}$ sec	T_{CP} = Central processor time
$T_{MJ} = 1 * 10^{-6}$ sec	I = Number of instructions executed
$T_{MT} = \begin{cases} 100 * T_{MJ} \text{ average} \\ 200 * T_{MJ} \text{ maximum} \end{cases}$	T_{ECT} = Extended core transfer time with access
$T_{EC} = (1.6 + 3.2) * 10^{-6}$ sec	T_{DT} = Disk transfer time
$T_{DL} = 33.5$ ms	
$T_D = 7.91$ µs	T_{ITRK} = Time to compute k planes
$T_{DP} = (3 * 67 + T_{DL})$ ms	$T_{I/O}$ = Time to set up disk I/O request
$P_O = 6$	T_{OV} = Overlap time of I/O and compute
$W_O = 495 * T_M$	
$I_O = (495 + 184 (P_O - 1)) * T_M$	K = 1
P = 100	

3. Formulas:

$$T_{CP} = T_M * P * \left[104 + (N_z - 2) \left(334 + (N_y - 2) \left(84 + (N_x - 2)(715) \right) \right) \right]$$

$$I = P * \left[20 + (N_z - 2) \left(32 + (\dot{N}_y - 2) \left(16 + (N_x - 2)(48) \right) \right) \right]$$

$$T_{DT} = P * \left[T_{DP} + T_{DL} + \frac{1}{2} * T_D * N_x * N_y * 9 \right] * (N_z)$$

$$T_{ECT} = P * 2 * \left[T_{EC} + T_M * N_x * N_y * 2 \right] * (N_z)$$

$$T_{ITRK} = \left[334 + (N_y - 2) \left(84 + (N_x - 2)(715) \right) \right] * T_M$$

$$T_{I/O} = P * \left[2 * (I_O + T_{MT} + W_O) \right] * (N_z)$$

$$T_{OV} = T_{ITRK}, \quad \text{where } 0 \leq T_{OV} \leq T_{DP}$$

$$T = T_{CP} + T_{ECT} + \left(T_{DT} - P * N_z * T_{ITRK} \right) + T_{I/O} + T_{MT}$$

30

4. Results:

N_x	N_y	N_z	T_{CP} (sec)	I	T_{ECT} (sec)	T_{DT} (sec)	$T_{I/O}$ (sec)	T_{ITRK} (sec)	T (sec)
100	100	100	6,737.918	453×10^7	40.096	6,239.500	5.988	.688	13,050.900

Summary on Problem No. 2A

6800 Times

N_x	N_y	N_z	T_{CP} (sec)	I	T_{ECT} (sec)	T_{DT} (sec)	$T_{I/O}$ (sec)	T_{ITRK} (sec)	T (sec)
25	25	5	.371	773×10^4	-	-	-	-	.371
25	25	25	2.842	593×10^5	.013	-	-	-	2.855
50	50	50	25.464	534×10^6	2.814	152.010	.035	.021	154.073
100	100	100	215.355	453×10^7	10.048	6,239.500	1.209	.022	6,245.612

6600 Times

N_x	N_y	N_z	T_{CP} (sec)	I	T_{ECT} (sec)	T_{DT} (sec)	$T_{I/O}$ (sec)	T_{ITRK} (sec)	T (sec)
25	25	5	1.484	773×10^4	-	-	-	-	1.484
25	25	25	11.369	593×10^5	.051	-	-	-	11.420
50	50	50	101.857	534×10^6	11.254	152.010	.145	.084	160.266
100	100	100	861.422	453×10^7	40.096	6239.500	4.836	.088	6,265.854

6400 Times

N_x	N_y	N_z	T_{CP} (sec)	I	T_{ECT} (sec)	T_{DT} (sec)	$T_{I/O}$ (sec)	T_{ITRK} (sec)	T (sec)
25	25	5	11.416	773×10^4	-	-	-	-	11.416
25	25	25	87.516	593×10^5	.051	-	-	-	87.567
50	50	50	792.829	534×10^6	11.254	152.010	.179	.661	812.709
100	100	100	6,737.918	453×10^7	40.096	6,239.500	5.988	.688	13,050.900

32

SOLUTION OF ELLIPTIC PARTIAL DIFFERENTIAL EQUATIONS
FIXED COEFFICIENTS

1. Obtain the time required to perform $P = 100$ iterations of the following:

$$U_{i,j,k}^{(P)} = W1 * U_{i+1,j,k}^{(P-1)} + W2 * U_{i-1,j,k}^{(P)} + W3 * U_{i,j+1,k}^{(P-1)}$$

$$+ W4 * U_{i,j-1,k}^{(P)} + W5 * U_{i,j,k+1}^{(P-1)} + W6 * U_{i,j,k-1}^{(P)}$$

$$+ W7 * U_{i,j,k}^{(P-1)} + W8 * F_{i,j,k}$$

for: $2 \leq i \leq n_x - 1$, $2 \leq j \leq n_y - 1$, $2 \leq k \leq n_z - 1$

$$d^P = \frac{1}{n_x \, n_y \, n_z} \sum_{k=2}^{n_z-1} \sum_{j=2}^{n_y-2} \sum_{i=2}^{n_x-2} \left| U_{i,j,k}^P - U_{i,j,k}^{(P-1)} \right|$$

2. Illustrative program: As in problem No. 2A.

3. CASE 1: All data is kept in core storage for
   a. $N_x = 25$, $N_y = 25$, $N_z = 5$
   b. $N_x = 25$, $N_y = 25$, $N_z = 25$

   CASE 2: All data is kept in extended core storage. Central memory is used as work area containing 25 planes of both U(I, J) and F(I, J). For 1 iteration 2 passes are necessary.

   CASE 3: U(I, J, K) is kept in extended core storage.
   F(I, J, K) is kept on the 2 disks.
   10 planes of F(I, J, K) can be kept on 1 position.
   10 positions on the disk are required to hold the necessary data.

1. Case No. 1:

$$(N_x = 25, \ N_y = 25, \ (N_z = 5, \ N_z = 25))$$

2. Notation

$T_M = 25 * 10^{-9} \ \text{sec}$ $T_{MJ} = 250 * 10^{-9} \ \text{sec}$ $P = 100$ $T_{MT} = \begin{cases} 100 * T_{MJ} \ \text{average} \\ 200 * T_{MJ} \ \text{maximum} \end{cases}$	T_{CP} = Central processor time I = Number of instructions executed T = Total job time

3. Formulas:

$$T_{CP} = T_M * P * \left[75 + (N_z - 2)\left(138 + (N_y - 2)\left(45 + (N_x - 2)(91) \right) \right) \right]$$

$$I = P * \left[20 + (N_z - 2)\left(32 + (N_y - 2)\left(14 + (N_x - 2)(48) \right) \right) \right]$$

$$T = T_{CP} + T_{MT}$$

4. Results:

N_x	N_y	N_z	T_{CP} (sec)	I	T (sec)
25	25	5	0.370	772×10^4	0.370
25	25	25	2.836	592×10^5	2.836

1. **Case No. 2:**

   $(N_x = 50, \ N_y = 50, \ N_z = 50)$

2. **Notation:**

$T_M = 25 * 10^{-9}$ sec $T_{MJ} = 250 * 10^{-9}$ sec $P = 100$ $T_{MT} = \begin{cases} 100 * T_{MJ} \text{ average} \\ 200 * T_{MJ} \text{ maximum} \end{cases}$ $T_{EC} = [.8 + 1.6] * 10^{-6}$ sec	T_{CP} = Central processor time I = Number of instructions executed T_{ECT} = Extended core transfer time with access T = Total job time

3. **Formulas:**

$$T_{CP} = T_M * P * \left[75 + (N_z-2)\left(138 + (N_y-2)\left(45 + (N_x-2)(91)\right)\right) \right]$$

$$I = P * \left[20 + (N_z-2)\left(32 + (N_y-2)\left(14 + (N_x-2)(48)\right)\right) \right]$$

$$T_{ECT} = 2 * P * [T_{EC} + T_M * N_x * N_y * 25]$$

$$T = T_{CP} + T_{ECT} + T_{MT}$$

4. **Results:**

N_x	N_y	N_z	T_{CP} (sec)	I	T_{ECT} (sec)	T (sec)
50	50	50	25.436	534×10^6	0.313	25.749

6800 Timing for Problem No. 2B

1. **Case No. 3:**

   $(N_x = 100, N_y = 100, N_z = 100)$

2. **Notation:**

$T_M = 25 * 10^{-9}$ sec	T_{CP} = Central processor time
$T_{MJ} = 250 * 10^{-9}$ sec	I = Number of instructions executed
$P = 100$	
$T_{MT} = \begin{cases} 100 * T_{MJ} \text{ average} \\ 200 * T_{MJ} \text{ maximum} \end{cases}$	T_{ECT} = Extended core transfer time with access
	T_{DT} = Disk transfer time
$T_{EC} = [.8 + 1.6] * 10^{-6}$ sec	T_{ITRK} = Time to compute k planes
$T_{DL} = 33.5$ ms	$K = 4$
$T_D = 7.91$ μs	$T_{I/O}$ = Time to set up disk I/O request
$T_{DP} = (3 * 67 + T_{DL})$ ms	
$W_o = 385 * T_M$	T_{OV} = Overlap time of I/O and compute
$I_o = [385 + 108(P_o-1)] * T_M$	
$P_o = 6$	

3. **Formulas**

$$T_{CP} = T_M * P * \left[75 + (N_z-2)\left(138 + (N_y-2)\left(45 + (N_x-2)(91)\right)\right)\right]$$

$$I = P * \left[20 + (N_z-2)\left(32 + (N_y-2)\left(14 + (N_x-2)(48)\right)\right)\right]$$

$$T_{ECT} = P * \left[T_{EC} + T_M * N_x * N_y * 4\right] * 25$$

$$T_{DT} = P * \left[\tfrac{1}{2} T_{DP} + T_{DL} + \tfrac{1}{2} T_D * N_x * N_y * 4\right] * 25$$

$$T_{ITRK} = k * \left[138 + (N_y-2)\left(45 + (N_x-2)(91)\right)\right] * T_M$$

Formulas (Cont.)

$$T_{I/O} = P * \left[2 * \left(I_O + T_{MT} + W_O \right) \right] * 25$$

$$T_{OV} = T_{ITRK}, \quad 0 \leq T_{OV} \leq \left[T_{DL} + \frac{1}{2} T_D * N_x * N_y \right]$$

$$T = T_{CP} + T_{ECT} + \left(T_{DT} - P * \left(\frac{N_z}{K} \right) * T_{ITRK} \right) + T_{I/O} + T_{MT}$$

4. Results

N_x	N_y	N_z	T_{CP} (sec)	I	T_{ECT} (sec)	T_{DT} (sec)	$T_{I/O}$ (sec)	T_{ITRK} (sec)	T (sec)
100	100	100	215.236	453×10^7	2.506	772.375	0.172	.088	769.849

6600 Timing for Problem No. 2B

1. Case No. 1:

$(N_x = 25,\ N_y = 25,\ (N_z = 5,\ N_z = 25))$

2. Notation:

$T_M = 100 * 10^{-9}$ sec $T_{MJ} = 1 * 10^{-6}$ sec $P = 100$	T_{CP} = Central processor time I = Number of instructions executed T = Total job time

3. Formulas

$$T_{CP} = T_M * P * \left[75 + (N_z-2)\Big(138 + (N_y-2)\big(45 + (N_x-2)(91)\big)\Big)\right]$$

$$I = P * \left[20 + (N_z-2)\Big(32 + (N_y-2)\big(14 + (N_x-2)(48)\big)\Big)\right]$$

$$T = T_{CP} + T_{MT}$$

4. Results

N_x	N_y	N_z	T_{CP} (sec)	I	T (sec)
25	25	5	1.480	$722 * 10^4$	1.480
25	25	25	11.343	$592 * 10^5$	11.343

6600 Timing for Problem No. 2B

1. Case No. 2:

    ($N_x = 50$, $N_y = 50$, $N_z = 50$)

2. Notation:

$T_M = 100 * 10^{-9}$ sec $T_{MJ} = 1 * 10^{-6}$ sec $P = 100$ $T_{MT} = \begin{array}{l} 100 * T_{MJ} \text{ average} \\ 200 * T_{MJ} \text{ maximum} \end{array}$ $T_{EC} = \left[1.6 + 3.2\right] * 10^{-6}$ sec	T_{CP} = Central processor time I = Number of instructions executed T_{ECT} = Extended core transfer time with access T = Total job time

3. Formulas:

$$T_{CP} = T_M * P * \left[75 + (N_z-2)\left(138 + (N_y-2)\left(45 + (N_x-2)\,(91)\right)\right)\right]$$

$$I = P * \left[20 + (N_z-2)\left(32 + (N_y-2)\left(14 + (N_x-2)\,(48)\right)\right)\right]$$

$$T_{ECT} = 2 * P * \left[T_{EC} + T_M * N_x * N_y * 25\right]$$

$$T = T_{CP} + T_{MT} + T_{ECT}$$

4. Results:

N_x	N_y	N_z	T_{CP} (sec)	I	T_{ECT} (sec)	T (sec)
50	50	50	101.743	534×10^6	1.251	102.992

6600 Timing for Problem No. 2B

1. Case No. 3:

   $(N_x = 100, N_y = 100, N_z = 100)$

2. Notation:

$T_M = 100 * 10^{-9}$ sec	T_{CP} = Central processor time
$T_{MJ} = 1 * 10^{-6}$ sec	I = Number of instructions executed
$P = 100$	
$T_{MT} = \begin{cases} 100 * T_{MJ} \text{ average} \\ 200 * T_{MJ} \text{ maximum} \end{cases}$	T_{ECT} = Extended core transfer time with access
$T_{EC} = \left[1.6 + 3.2\right] * 10^{-6}$ sec	T_{DT} = Disk transfer time
$T_{DL} = 33.5$ ms	T_{ITRK} = Time to compute k planes
$T_D = 7.91$ μs	$K = 4$
$T_{DP} = \left(3 * 67 + T_{DL}\right)$ ms	$T_{I/O}$ = Time to set up disk I/O request
$W_o = 385 * T_M$	T_{OV} = Overlap time of I/O and compute
$I_o = \left[385 + 108 (P_o-1)\right] * T_M$	
$P_o = 6$	

3. Formulas:

$$T_{CP} = T_M * P * \left[75 + (N_z-2)\left(138 + (N_y-2)\left(45 + (N_x-2)(91)\right)\right)\right]$$

$$I = P * \left[20 + (N_z-2)\left(32 + (N_y-2)\left(14 + (N_x-2)(48)\right)\right)\right]$$

$$T_{ECT} = P * \left[T_{EC} + T_M * N_x * N_y * 4\right] * 25$$

$$T_{DT} = P * \left[\frac{1}{2} T_{DP} + T_{DL} + \frac{1}{2} T_D * N_x * N_y * 5\right] * 20$$

$$T_{ITRK} = K * \left[138 + (N_y-2)\left(45 + (N_x-2)(91)\right)\right] * T_M$$

$$T_{I/O} = P * 2 * \left[I_o + T_{MT} + W_o\right] * 20$$

$$T_{OV} = T_{ITRK}, \quad 0 \le T_{OV} \le \left[T_{DL} + \frac{1}{2} T_{DP} + \frac{1}{2} T_D * N_x * N_y * 4\right]$$

$$T = T_{CP} + T_{ECT} + (T_{DT} - P * T_{ITRK}) + T_{I/O} + T_{MT}$$

4. Results:

N_x	N_y	N_z	T_{CP} (sec)	I	T_{ECT} (sec)	T_{DT} (sec)	$T_{I/O}$ (sec)	T_{ITRK} (sec)	T (sec)
100	100	100	860. 943	453×10^7	10. 012	772. 375	. 689	. 352	870. 060

6400 Timing for Problem No. 2B

1. Case No. 1: $(N_x = 25, N_y = 25, (N_z = 5, N_z = 25))$

2. Notation:

$T_M = 100 * 10^{-9}$ sec $T_{MJ} = 1 * 10^{-6}$ sec $P = 100$ $T_{MT} = \begin{cases} 100 * T_{MJ} \text{ average} \\ 200 * T_{MJ} \text{ maximum} \end{cases}$	T_{CP} = Central processor time I = Number of instructions executed T = Total job time

3. Formulas:

$$T_{CP} = T_M * P * \left[104 + (N_z-2)\left(334 + (N_y-2)\left(74 + (N_x-2)(715)\right)\right) \right]$$
$$I = P * \left[20 + (N_z-2)\left(32 + (N_y-2)\left(14 + (N_x-2)(48)\right)\right) \right]$$
$$T = T_{CP} + T_{MT}$$

4. Results:

N_x	N_y	N_z	T_{CP} (sec)	I	T (sec)
25	25	5	11.409	772×10^4	11.409
25	25	25	87.463	592×10^5	87.463

6400 Timing for Problem No. 2B

1. Case No. 2: $(N_x = 50, N_y = 50, N_z = 50)$

2. Notation:

$T_M = 100 * 10^{-9}$ sec $T_{MJ} = 1 * 10^{-6}$ sec $P = 100$ $T_{MT} = \begin{cases} 100 * T_{MJ} \text{ average} \\ 200 * T_{MJ} \text{ maximum} \end{cases}$ $T_{EC} = \lfloor 1.6 + 3.2 \rfloor * 10^{-6}$ sec	T_{CP} = Central processor time I = Number of instructions executed T_{ECT} = Extended core transfer time with access T = Total job time

3. Formulas:

$$T_{CP} = T_M * P * \left[104 + (N_z - 2)\Big(334 + (N_y - 2)\big(74 + (N_x - 2)(715)\big)\Big) \right]$$

$$I = P * \left[20 + (N_z - 2)\Big(32 + (N_y - 2)\big(14 + (N_x - 2)(48)\big)\Big) \right]$$

$$T_{ECT} = 2 * P * \left[T_{EC} + T_M * N_x * N_y * 25 \right]$$

$$T = T_{CP} + T_{ECT} + T_{MT}$$

4. Results:

N_x	N_y	N_z	T_{CP} (sec)	I	T_{ECT} (sec)	T (sec)
50	50	50	792.599	534×10^6	1.251	793.850

6400 Timing for Problem No. 2B

1. Case No. 3: ($N_x = 100$, $N_y = 100$, $N_z = 100$)

2. Notation:

$$T_M = 100 * 10^{-9} \text{ sec}$$

$$T_{MJ} = 1 * 10^{-6} \text{ sec}$$

$$P = 100$$

$$T_{MT} = \begin{cases} 100 * T_{MJ} \text{ average} \\ 200 * T_{MJ} \text{ maximum} \end{cases}$$

$$T_{EC} = \left[1.6 + 3.2\right] * 10^{-6} \text{ sec}$$

$$T_{DL} = 33.5 \text{ ms}$$

$$T_D = 7.91 \text{ μs}$$

$$T_{DP} = \left(3 * 67 + T_{DL}\right) \text{ms}$$

$$W_o = 495 * T_M$$

$$I_o = \left[495 + 184\,(P_o - 1)\right] * T_M$$

$$P_o = 6$$

T_{CP} = Central processor time

I = Number of instructions executed

T_{ECT} = Extended core transfer time with access

T_{DT} = Disk transfer time

T_{ITRK} = Time to compute k planes

$K = 1$

$T_{I/O}$ = Time to set up disk I/O request

T_{OV} = Overlap time of I/O and compute

3. Formulas:

$$T_{CP} = T_M * P * \left[104 + (N_z - 2)\left(334 + (N_y - 2)\left(74 + (N_x - 2)(715)\right)\right)\right]$$

$$I = P * \left[20 + (N_z - 2)\left(32 + (N_y - 2)\left(14 + (N_x - 2)(48)\right)\right)\right]$$

$$T_{ECT} = P * \left[T_{EC} + T_M * N_x * N_y * 5\right] * 25$$

$$T_{DT} = P * \left[\frac{1}{2} T_{DP} + T_{DL} + \frac{1}{2} T_D * N_x * N_y * 4\right] * 25$$

$$T_{ITRK} = K * \left[334 + (N_y - 2)\left(74 + (N_y - 2)(715)\right)\right] * T_M$$

$$T_{I/O} = P * \left[2 * (I_o + T_{MT} + W_o)\right] * 20$$

$$T_{OV} = T_{ITRK}, \quad 0 \leq T_{OV} \leq \left[T_{DL} + \frac{1}{2} T_{DP} + \frac{1}{2} T_D * N_x * N_y * 4\right]$$

$$T = T_{CP} + T_{ECT} + (T_{DT} - P * T_{ITRK}) + T_{I/O} + T_{MT}$$

4. Results:

N_x	N_y	N_z	T_{CP} (sec)	I	T_{ECT} (sec)	T_{DT} (sec)	$T_{I/O}$ (sec)	T_{ITRK} (sec)	T (sec)
100	100	100	6, 736. 958	453×10^7	10. 012	772. 375	1. 164	2. 750	6, 748. 134

Summary of Problem No. 2B

6800 Times

N_x	N_y	N_z	T_{CP} (sec)	I	T_{ECT} (sec)	T_{DT} (sec)	$T_{I/O}$ (sec)	T_{ITRK} (sec)	T (sec)
25	25	5	0. 370	772×10^4	-	-	-	-	0. 370
25	25	25	2. 836	592×10^5	-	-	-	-	2. 836
50	50	50	25. 436	534×10^6	0. 313	-	-	-	25. 749
100	100	100	215. 236	453×10^7	2. 506	772. 375	0. 172	0. 088	769. 849

6600 Times

N_x	N_y	N_z	T_{CP} (sec)	I	T_{ECT} (sec)	T_{DT} (sec)	$T_{I/O}$ (sec)	T_{ITRK} (sec)	T (sec)
25	25	5	1. 480	722×10^4	-	-	-	-	1. 480
25	25	25	11. 343	592×10^5	-	-	-	-	11. 343
50	50	50	101. 743	534×10^6	1. 251	-	-	-	102. 992
100	100	100	860. 943	453×10^7	10. 012	772. 375	0. 689	0. 352	870. 060

6400 Times

N_x	N_y	N_z	T_{CP} (sec)	I	T_{ECT} (sec)	T_{DT} (sec)	$T_{I/O}$ (sec)	T_{ITRK} (sec)	T (sec)
25	25	5	11. 409	772×10^4	-	-	-	-	11. 409
25	25	25	87. 463	592×10^5	-	-	-	-	87. 463
50	50	50	792. 599	534×10^6	1. 251	-	-	-	793. 850
100	100	100	6, 736. 958	453×10^7	10. 012	772. 375	1. 164	2. 750	6, 748. 134

PROBLEM NO. 3A
MATRIX MULTIPLICATION

1. For $n = 100\ (100)\ 500$, find the time to obtain

$$C_{ij} = \sum_{k=1}^{n} a_{ik}\, b_{kj}$$

   for: $1 \le i ,\ j \le n$

2. Illustrative Problem:

        DO 1 I = 1, N
        DO 1 J = 1, N
        C(I, J) = 0, 0
        DO 1 K = 1, N
   1   C(I, J) = C(I, J) + A(I, K) * B(I, K)

3. Two cases are to be considered:

   a. All data is kept in core storage for N = 100, 200.

   b. All data is kept in extended core for n = 300, 400, 500 and core storage is used as work area.

6800 Timing for Problem No. 3A

1. Case No. 1: (n = 100, 200)

2. Notation:

$$T_M = 25 * 10^{-9} \text{ sec}$$

$$T_{MJ} = 250 * 10^{-9} \text{ sec}$$

$$T_{MT} = \begin{cases} 100 * T_{MJ} \text{ average} \\ 200 * T_{MJ} \text{ maximum} \end{cases}$$

T_{CP} = Central processor time

I = Number of instructions executed

T = Total job time

3. Formulas:

$$T_{CP} = T_M * \left[32 + N \left(16 + N \left(23 + \left(\frac{N+2}{2}\right)(18) \right) \right) \right]$$

$$I = \left[10 + N \left(5 + N \left(8 + \left(\frac{N+2}{2}\right)(10) \right) \right) \right]$$

$$T = T_{CP} + T_{MT}$$

4. Results:

N	T_{CP} (sec)	I	T (sec)
100	0.235	518×10^4	0.235
200	1.841	407×10^5	1.841

48

6800 Timing for Problem No. 3A

1. Case No. 2: (N = 300, 400, 500)

2. Notation:

$T_M = 25 * 10^{-9}$ sec $T_{MJ} = 250 * 10^{-9}$ sec $T_{MT} = \begin{cases} 100 * T_{MJ} \text{ average} \\ 200 * T_{MJ} \text{ maximum} \end{cases}$ $T_{EC} = [0.8 + 1.6] * 10^{-6}$ sec	T_{CP} = Control processor time I = Number of instructions executed T_{ECT} = Extended core transfer time with access $B = \begin{cases} 100 \text{ when N = 300} \\ 100 \text{ when N = 400} \\ 80 \text{ when N = 500} \end{cases}$

3. Formulas:

$$T_{CP} = T_M * \left[32 + N\left(16 + N\left(23 + \left(\frac{N+2}{2}\right)(18)\right)\right)\right]$$

$$I = \left[10 + N\left(5 + N\left(8 + \left(\frac{N+2}{2}\right)(10)\right)\right)\right]$$

$$T_{ECT} = \left[T_{EC} + 2*T_M*N*B\right] * \left(\frac{N-B}{B}\right)^2 + \left[T_{EC} + T_M*N*B\right]\frac{N}{B}$$

$$T = T_{CP} + T_{ECT} + T_{MT}$$

4. Results:

N	T_{CP} (sec)	I	T_{ECT} (sec)	T (sec)
300	6.167	136×10^6	0.008	6.175
400	14.564	322×10^6	0.022	14.586
500	28.381	624×10^6	0.061	28.442

6600 Timing for Problem No. 3A

1. Case No. 1: (n = 100, 200)

2. Notation:

$T_M = 100 * 10^{-9}$ sec $T_{MJ} = 1 * 10^{-6}$ sec $T_{MT} = \begin{cases} 100 * T_{MJ} \text{ average} \\ 200 * T_{MJ} \text{ maximum} \end{cases}$	T_{CP} = Central processor time I = Number of instructions executed T = Total job time

3. Formulas:

$$T_{CP} = T_M * \left[32 + N \left(16 + N \left(23 + \left(\frac{N+2}{2}\right)(18) \right) \right) \right]$$

$$I = \left[10 + N \left(5 + N \left(8 + \left(\frac{N+2}{2}\right)(10) \right) \right) \right]$$

$$T = T_{CP} + T_{MT}$$

4. Results:

N	T_{CP} (sec)	I	T (sec)
100	0.941	518×10^4	0.941
200	7.364	407×10^5	7.364

6600 Timing for Problem No. 3A

1. Case No. 2: (N = 300, 400, 500)

2. Notation:

$T_M = 100 * 10^{-9}$ sec $T_{MJ} = 1 * 10^{-6}$ sec $T_{MT} = \begin{cases} 100 * T_{MJ} \text{ average} \\ 200 * T_{MJ} \text{ maximum} \end{cases}$ $T_{EC} = \left[1.6 + 3.2\right] * 10^{-6}$ sec	T_{CP} = Central processor time I = Number of instructions executed T_{ECT} = Extended core transfer time with access T = Total job time $B = \begin{cases} 100 \text{ when N = 300} \\ 100 \text{ when N = 400} \\ 80 \text{ when N = 500} \end{cases}$

3. Formulas:

$$T_{CP} = T_M * \left[32 + N\left(16 + N\left(23 + \left(\frac{N+2}{2}\right)(18)\right)\right)\right]$$

$$I = \left[10 + N\left(5 + N\left(8 + \left(\frac{N+2}{2}\right)(10)\right)\right)\right]$$

$$T_{ECT} = \left[T_{EC} + 2 * T_M * N * B\right] * \left(\frac{N-B}{B}\right)^2 + \left[T_{EC} + T_M * N * B\right] * \frac{N}{B}$$

$$T = T_{CP} + T_{ECT} + T_{MT}$$

4. Results:

N	T_{CP} (sec)	I	T_{ECT} (sec)	T (sec)
300	24.669	136×10^6	0.033	24.702
400	58.257	322×10^6	0.088	58.345
500	113.529	624×10^6	0.245	113.774

6400 Timing for Problem No. 3A

1. Case No. 1: (N = 100, 200)

2. Notation:

$T_M = 100 * 10^{-9}$ sec $T_{MJ} = 1 * 10^{-6}$ sec $T_{MT} = \begin{cases} 100 * T_{MJ} \text{ average} \\ 200 * T_{MJ} \text{ maximum} \end{cases}$	T_{CP} = Central processor time I = Number of instructions executed T = Total job time

3. Formulas:

$$T_{CP} = T_M * \left[50 + N \left(31 + N \left(52 + \left(\frac{N+2}{2}\right)(173) \right) \right) \right]$$

$$I = \left[10 + N \left(5 + N \left(8 + \left(\frac{N+2}{2}\right)(10) \right) \right) \right]$$

$$T = T_{CP} + T_{MT}$$

4. Results:

N	T_{CP} (sec)	I	T (sec)
100	8. 875	518×10^4	8. 875
200	70. 101	407×10^5	70. 101

6400 Timing for Problem No. 3A

1. Case No. 2: (N = 300, 400, 500)

2. Notation:

$$T_M = 100 * 10^{-9} \text{ sec}$$

$$T_{MJ} = 1 * 10^{-6} \text{ sec}$$

$$T_{MT} = \begin{cases} 100 * T_{MJ} \text{ average} \\ 200 * T_{MJ} \text{ maximum} \end{cases}$$

$$T_{EC} = \left[1.6 + 3.2\right] * 10^{-6} \text{ sec}$$

T_{CP} = Control processor time

I = Number of instructions executed

T_{ECT} = Extended core transfer time with access

T = Total job time

$$B = \begin{cases} 100 \text{ when } N = 300 \\ 100 \text{ when } N = 400 \\ 80 \text{ when } N = 500 \end{cases}$$

3. Formulas:

$$T_{CP} = T_M * \left[50 + N\left(31 + N\left(52 + \left(\tfrac{N+2}{2}\right)(173)\right)\right)\right]$$

$$I = \left[10 + N\left(5 + N\left(8 + \left(\tfrac{N+2}{2}\right)(10)\right)\right)\right]$$

$$T_{ECT} = \left[T_{EC} + 2 * T_M * N * B\right] * \left(\tfrac{N-B}{B}\right)^2 + \left[T_{EC} + T_M * N * B\right]\tfrac{N}{B}$$

$$T = T_{CP} + T_{ECT} + T_{MT}$$

4. Results:

N	T_{CP} (sec)	I	T_{ECT} (sec)	T (sec)
300	235.576	136×10^6	0.033	235.609
400	557.201	322×10^6	0.088	557.289
500	1086.877	624×10^6	0.245	1087.122

Summary Problem 3A

6800 Timing

N	T_{CP} (sec)	I	T_{ECT} (sec)	T (sec)
100	0. 235	518×10^4	—	0. 235
200	1. 841	407×10^5	—	1. 841
300	6. 167	136×10^6	0. 008	6. 175
400	14. 564	322×10^6	0. 022	14. 586
500	23. 381	624×10^6	0. 061	28. 442

6600 Timings

N	T_{CP} (sec)	I	T_{ECT} (sec)	T (sec)
100	0. 941	518×10^4	—	0. 941
200	7. 364	407×10^5	—	7. 364
300	24. 669	136×10^6	0. 033	24. 702
400	58. 257	322×10^6	0. 088	58. 345
500	113. 529	624×10^6	0. 245	113. 774

6400 Timings

N	T_{CP} (sec)	I	T_{ECT} (sec)	T (sec)
100	8. 875	518×10^4	—	8. 875
200	70. 101	407×10^5	—	70. 101
300.	235. 576	136×10^6	0. 033	235. 609
400	557. 201	322×10^6	0. 088	557. 289
500	1086. 877	624×10^6	0. 245	1087. 122

SOLUTION OF LINEAR EQUATIONS WITH N RIGHT-HAND SIDES

1. Find the time required to solve the n x n system of equations when the matrix B has dimensions n x 1, for n = 100, (100), 500.

2. Illustrative Program:

```
         N = NN
         M = MM
         N1 = N-1
         DO 9  J = 1, N1
         L = J + 1
6        DO 9  I = L, N
         R = - A(I, J)/A(J, J)
         DO 7  K = 2, N
7        A(I, K) = A(I, K) + R * A(J, K)
         DO 8  K = 1, N
8        B(I, K) = B(I, K) + R * B(J, K)
9        CONTINUE
         ERR = 1. 0
         DO 11  I = 1, N
         IF(A (I, I)) 11, 10, 11
10       ERR = -1. 0
         GO TO 15
11       CONTINUE
         DO 14  K = 1, M
         B(N, K) = B(N, K)/A(N, N)
         DO 13  L = 1, N1
         I = N - L
         R = 0. 0
         IMAX = I + 1
         DO 12  J = IMAX, N
12       R = R + A(I, J) * B(J, K)
13       B(I, K) = (B (I, K) - R)/A(I, I)
14       CONTINUE
15       RETURN
```

3. Two cases are timed:

    a. All data is kept in core storage.

    b. All data is kept in extended core. All of core is used as work area.

1. Case No. 1: (n = 100, 200, 300)

2. Notation:

$T_M = 25 * 10^{-9}$ sec $T_{MJ} = 250 * 10^{-9}$ sec $T_{MT} = \begin{cases} 100 * T_{MJ} \text{ average} \\ 200 * T_{MJ} \text{ maximum} \end{cases}$	T_{CP} = Central processor time I = Number of instructions executed T = Total job time

3. Formulas:

$$T_{CP} = T_M * \left[136 + (N-1) \left(51 + \left(\frac{N+1}{2}\right) \left(55 + \left(\frac{N+1}{2}\right)\left(\frac{23}{2}\right)\right)\right)\right.$$
$$\left. + N \left(73 + (N-1) \left(38 + \left(\frac{N+1}{2}\right)\left(\frac{18}{2}\right)\right)\right)\right]$$

$$I = \left[52 + (N-1) \left(17 + \left(\frac{N+1}{2}\right)\left(18 + \left(\frac{N+1}{2}\right)\left(\frac{12}{2}\right)\right)\right)\right.$$
$$\left. + N \left(22 + (N-1) \left(12 + \left(\frac{N+1}{2}\right)\left(\frac{10}{2}\right)\right)\right)\right]$$

$$T = T_{CP} + T_{MT}$$

4. Results:

N	T_{CP} (sec)	I	T (sec)
100	0. 202	422×10^4	0. 202
200	1. 543	329×10^5	1. 543
300	5. 133	110×10^6	5. 133

6800 Timing for Problem No. 3B

1. Case No. 2: (N = 400, 500)

2. Notation:

$T_M = 25 * 10^{-9}$ sec $T_{MJ} = 250 * 10^{-9}$ sec $T_{MT} = \begin{cases} 100 * T_{MJ} \text{ average} \\ 200 * T_{MJ} \text{ maximum} \end{cases}$ $T_{EC} = \lvert 0.8 + 1.6 \rvert * 10^{-6}$ sec	T_{CP} = Control processor time I = Number of instructions executed $\bullet$ T_{ECT} = Extended core transfer time with access T = Total Job Time $B = \begin{cases} 250 \text{ when N = 400} \\ 200 \text{ when N = 500} \end{cases}$

3. Formulas:

$$T_{CP} = T_M * \left[136 + (N-1)\left(51 + \left(\tfrac{N+1}{2}\right)\left(55 + \left(\tfrac{N+1}{2}\right)\left(\tfrac{23}{2}\right)\right)\right)\right.$$
$$\left. + N\left(73 + (N-1)\left(38 + \left(\tfrac{N+1}{2}\right)\left(\tfrac{18}{2}\right)\right)\right)\right]$$

$$I = \left[52 + (N-1)\left(17 + \left(\tfrac{N+1}{2}\right)\left(18 + \left(\tfrac{N+1}{2}\right)\left(\tfrac{12}{2}\right)\right)\right)\right.$$
$$\left. + N\left(22 + (N-1)\left(12 + \left(\tfrac{N+1}{2}\right)\left(\tfrac{10}{2}\right)\right)\right)\right]$$

$$T_{ECT} = 2 * \left[T_{EC} + T_M * N * B\right]\left(\tfrac{N}{B}\right)^2$$

$$T = T_{CP} + T_{ECT} + T_{MT}$$

4. Results

N	T_{CP} (sec)	I	T_{ECT} (sec)	T (sec)
400	12.074	259×10^6	.007	12.081
500	23.470	505×10^6	.026	23.493

6600 Timing for Problem No. 3B

1. Case No. 1: (N = 100, 200, 300)

2. Notation:

$T_M = 100 * 10^{-9}$ sec $T_{MJ} = 1 * 10^{-6}$ sec $T_{MT} = \begin{cases} 100 * T_{MJ} \text{ average} \\ 200 * T_{MJ} \text{ maximum} \end{cases}$	T_{CP} = Central processor time I = Number of instructions executed T = Total job time

3. Formulas:

$$T_{CP} = T_M * \left[136 + (N-1)\left(51 + \left(\tfrac{N+1}{2}\right)\left(55 + \left(\tfrac{N+1}{2}\right)\left(\tfrac{23}{2}\right)\right)\right) \right.$$
$$\left. + N\left(73 + (N-1)\left(38 + \left(\tfrac{N+1}{2}\right)\left(\tfrac{18}{2}\right)\right)\right) \right]$$

$$I = \left[52 + (N-1)\left(17 + \left(\tfrac{N+1}{2}\right)\left(18 + \left(\tfrac{N+1}{2}\right)\tfrac{12}{2}\right)\right) \right.$$
$$\left. + N\left(22 + (N-1)\left(12 + \left(\tfrac{N+1}{2}\right)\left(\tfrac{10}{2}\right)\right)\right) \right]$$

$$T = T_{CP} + T_{MT}$$

4. Results:

N	T_{CP} (sec)	I	T (sec)
100	0.806	422×10^4	0.806
200	6.173	329×10^5	6.173
300	20.530	110×10^6	20.530

6600 Timing for Problem No. 3B

1. Case No. 2: (N = 400, 500)

2. Notation:

$$T_M = 100 \times 10^{-9} \text{ sec}$$

$$T_{MJ} = 1 * 10^{-6} \text{ sec}$$

$$100 * T_{MJ} \text{ average}$$
$$200 * T_{MJ} \text{ maximum}$$

$$T_{EC} = \left[1.6 \times 3.2\right] * 10^{-6} \text{ sec}$$

T_{CP} = Control processor time

I = Number of instructions executed

T_{ECT} = Extended core transfer time with access

$$B = \begin{cases} 250 \text{ when } N = 400 \\ 200 \text{ when } N = 500 \end{cases}$$

3. Formulas:

$$T_{CP} = T_M * \left[136 + (N-1)\left(51 + \left(\tfrac{N+1}{2}\right)\left(55 + \left(\tfrac{N+1}{2}\right)\left(\tfrac{23}{2}\right)\right)\right)\right.$$
$$\left. + N\left(73 + (N-1)\left(38 + \left(\tfrac{N+1}{2}\right)\left(\tfrac{18}{2}\right)\right)\right)\right]$$

$$I = \left[52 + (N-1)\left(17 + \left(\tfrac{N+1}{2}\right)\left(18 + \left(\tfrac{N+1}{2}\right)\left(\tfrac{12}{2}\right)\right)\right)\right.$$
$$\left. + N\left(22 + (N-1)\left(12 + \left(\tfrac{N+1}{2}\right)\left(\tfrac{10}{2}\right)\right)\right)\right]$$

$$T_{ECT} = 2 * \left[T_{EC} + T_M * N * B\right]\left(\tfrac{N}{B}\right)^2$$

$$T = T_{CP} + T_{ECT} + T_{MT}$$

4. Results:

N	T_{CP} (sec)	I	T_{ECT} (sec)	T (sec)
400	48.297	259×10^6	.032	48.329
500	93.879	505×10^6	.125	94.004

6400 Timing for Problem No. 3B

1. Case No. 1: (n = 100, 200, 300)

2. Notation:

$T_M = 100 * 10^{-9}$ sec $T_{MJ} = 1 * 10^{-6}$ sec $T_{MT} = \begin{cases} 100 * T_{MJ} \text{ average} \\ 200 * T_{MJ} \text{ maximum} \end{cases}$	T_{CP} = Central processor time I = Number of instructions executed T = Total job time

3. Formulas:

$$T_{CP} = T_M * \left[268 + (N-1) \left(92 + \left(\tfrac{N+1}{2}\right) \left(209 + \left(\tfrac{N+1}{2}\right)\left(\tfrac{180}{2}\right)\right)\right) \right.$$

$$\left. + N \left(184 + (N-1) \left(145 + \left(\tfrac{N+1}{2}\right)\left(\tfrac{173}{2}\right)\right)\right) \right]$$

$$I = \left[52 + (N-1) \left(17 + \left(\tfrac{N+1}{2}\right)\left(18 + \left(\tfrac{N+1}{2}\right)\left(\tfrac{12}{2}\right)\right)\right) \right.$$

$$\left. + N \left(22 + (N-1) \left(12 + \left(\tfrac{N+1}{2}\right)\left(\tfrac{10}{2}\right)\right)\right) \right]$$

$$T = T_{CP} + T_{MT}$$

4. Results:

N	T_{CP} (sec)	I	T (sec)
100	6. 848	422×10^4	6. 848
200	53. 689	329×10^5	53. 689
300	179. 975	110×10^6	179. 975

1. Case No. 2: (N = 400, 500)

2. Notation:

$$T_M = 100 * 10^{-9} \text{ sec}$$

$$T_{MJ} = 1 * 10^{-6} \text{ sec}$$

$$T_{MT} = \begin{cases} 100 * T_{MJ} \text{ average} \\ 200 * T_{MJ} \text{ maximum} \end{cases}$$

$$T_{EC} = \left[1.6 \times 3.2 \right] * 10^{-6} \text{ sec}$$

T_{CP} = Control processor time

I = Number of instructions executed

T_{ECT} = Extended core transfer time with access

T = Total Job Time

$$B = \begin{cases} 250 \text{ when } N = 400 \\ 200 \text{ when } N = 500 \end{cases}$$

3. Formulas:

$$T_{CP} = T_M * \left[268 + (N-1) \left(92 + \left(\frac{N+1}{2}\right)\left(209 + \left(\frac{N+1}{2}\right)\left(\frac{180}{2}\right)\right)\right) \right.$$

$$\left. + N \left(184 + (N-1) \left(145 + \left(\frac{N+1}{2}\right)\left(\frac{173}{2}\right)\right)\right) \right]$$

$$I = \left[52 + (N-1) \left(17 + \left(\frac{N+1}{2}\right)\left(18 + \left(\frac{N+1}{2}\right)\left(\frac{12}{2}\right)\right)\right) \right.$$

$$\left. + N \left(22 + (N-1) \left(12 + \left(\frac{N+1}{2}\right)\left(\frac{10}{2}\right)\right)\right) \right]$$

$$T_{ECT} = 2 * \left[T_{EC} + T_M * N * B \right] \left(\frac{N}{B}\right)^2$$

$$T = T_{CP} + T_{ECT} + T_{MT}$$

4. Results:

N	T_{CP} (sec)	I	T_{ECT} (sec)	T (sec)
400	412.150	.259 x 10^6	.032	412.180
500	828.670	505 x 10^6	.125	828.800

Summary Problem 3B

6800 Timings

N	T_{CP} (sec)	I	T_{ECT} (sec)	T (sec)
100	0.202	422×10^4	—	0.202
200	1.543	329×10^5	—	1.543
300	5.133	110×10^6	—	5.133
400	12.074	259×10^6	.007	12.081
500	23.470	505×10^6	.026	23.493

6600 Timings

N	T_{CP} (sec)	I	T_{ECT} (sec)	T (sec)
100	0.806	422×10^4	—	0.806
200	6.173	329×10^5	—	6.173
300	20.530	110×10^6	—	20.530
400	48.297	259×10^6	.032	48.329
500	93.879	505×10^6	.125	94.004

6400 Timings

N	T_{CP} (sec)	I	T_{ECT} (sec)	T (sec)
100	6.848	422×10^4	—	6.848
200	53.689	329×10^5	—	53.689
300	179.975	110×10^6	—	179.975
400	412.15	259×10^6	.032	412.18
500	828.67	505×10^6	.125	828.80

PROBLEM NO. 3C

SOLUTION OF LINEAR EQUATIONS WITH N x N RIGHT-HAND SIDES

1. Find the time required to solve the n x n system of equations when the matrix B has dimensions n x n for n = 100, (100), 500

2. Illustrative program as in Problem No. 3B.

3. There are two cases to be considered:

    a. All data is kept in core: (n = 100, 200)

    b. All data is kept in extended core: (n = 300, 400, 500).
        Central memory is used for intermediate results only.

6800 Timing for Problem No. 3C

1. Case No. 1: (N = 100, 200)

2. Notation:

$T_M = 25 * 10^{-9}$ sec $T_{MJ} = 250 * 10^{-9}$ sec $T_{MT} = \begin{cases} 100 * T_{MJ} \text{ average} \\ 200 * T_{MJ} \text{ maximum} \end{cases}$	T_{CP} = Central processor time I = Number of instructions executed T = Total job time

3. Formulas:

$$T_{CP} = T_M \left[136 + (N-1) \left(51 + \left(\frac{N+1}{2} \right) \left(43 + \left(\frac{N+1}{2} \right) \left(\frac{23}{2} \right) + N \left(\frac{23}{2} \right) \right) \right) \right.$$

$$\left. + N \left(73 + (N-1) \left(38 + \left(\frac{N+1}{2} \right) \left(\frac{18}{2} \right) \right) \right) \right]$$

$$I = \left[52 + (N-1) \left(17 + \left(\frac{N+1}{2} \right) \left(13 + \left(\frac{N+1}{2} \right) \left(\frac{12}{2} \right) \right) \right) \right.$$

$$\left. + N \left(22 + (N-1) \left(12 + \left(\frac{N+1}{2} \right) \left(\frac{10}{2} \right) \right) \right) \right]$$

$$T = T_{CP} + T_{MT}$$

4. Results:

N	T_{CP} (sec)	I	T (sec)
100	0. 344	720×10^4	0. 344
200	2. 683	568×10^5	2. 683

6800 Timing for Problem No. 3C

1. Case No. 2: $(N = 300, 400, 500)$

2. Notation:

$$T_M = 25 * 10^{-9} \text{ sec}$$

$$T_{MJ} = 250 * 10^{-9} \text{ sec}$$

$$T_{MT} = \begin{cases} 100 * T_{MJ} \text{ average} \\ 200 * T_{MJ} \text{ maximum} \end{cases}$$

$$T_{EC} = \left| 0.8 + 1.6 \right| * 10^{-6} \text{ sec}$$

T_{CP} = Control processor time

I = Number of instructions executed

T_{ECT} = Extended core transfer time with access

$$B = \begin{cases} 200 \text{ when } N = 300 \\ 150 \text{ when } N = 400 \\ 100 \text{ when } N = 500 \end{cases}$$

3. Formulas:

$$T_{CP} = T_M * \left[136 + (N-1) \left(51 + \left(\frac{N+1}{2}\right)\left(43 + \left(\frac{N+1}{2}\right)\left(\frac{23}{2}\right) + N \left(\frac{23}{2}\right)\right)\right) \right.$$

$$\left. + N \left(73 + (N-1) \left(38 + \left(\frac{N+1}{2}\right)\left(\frac{18}{2}\right)\right)\right) \right]$$

$$I = \left[52 + (N-1) \left(17 + \left(\frac{N+1}{2}\right)\left(13 + \left(\frac{N+1}{2}\right)\left(\frac{12}{2}\right) + N \left(\frac{12}{2}\right)\right)\right) \right.$$

$$\left. + N \left(22 + (N-1) \left(12 + \left(\frac{N+1}{2}\right)\left(\frac{10}{2}\right)\right)\right) \right]$$

$$T_{ECT} = 4 * \left[T_{EC} + \left(T_M * N * B\right) \right] \left(\frac{N-B}{B}\right) \left(\frac{N}{B}\right)$$

$$T = T_{CP} + T_{ECT} + T_{MT}$$

4. Results:

	T_{CP} (sec)	I	T_{ECT} (sec)	T (sec)
300	8.998	190×10^6	0.0045	9.002
400	21.249	451×10^6	0.0383	21.278
500	41.396	880×10^6	0.1002	41.496

6600 Timing for Problem No. 3C

1. Case No. 1: (N = 100, 200)

2. Notation:

$T_M = 100 * 10^{-9}$ sec $T_{MJ} = 1 * 10^{-6}$ sec $T_{MT} = \begin{cases} 100 * T_{MJ} \text{ average} \\ 200 * T_{MJ} \text{ maximum} \end{cases}$	T_{CP} = Central processor time I = Number of instructions executed T = Total job time

3. Formulas:

$$T_{CP} = T_M * \left[136 + (N-1) \left(51 + \left(\frac{N+1}{2}\right) \left(43 + \left(\frac{N+1}{2}\right)\left(\frac{23}{2}\right) + N\left(\frac{23}{2}\right)\right)\right)\right.$$

$$\left. + N \left(73 + (N-1) \left(38 + \left(\frac{N+1}{2}\right)\left(\frac{18}{2}\right)\right)\right)\right]$$

$$I = \left[52 + (N-1) \left(17 + \left(\frac{N+1}{2}\right) \left(13 + \left(\frac{N+1}{2}\right)\left(\frac{12}{2}\right) + N\left(\frac{12}{2}\right)\right)\right)\right.$$

$$\left. + N \left(22 + (N-1) \left(12 + \left(\frac{N+1}{2}\right)\left(\frac{10}{2}\right)\right)\right)\right]$$

$$T = T_{CP} + T_{MT}$$

4. Results:

N	T_{CP} (sec)	I	T (sec)
100	1.375	720×10^4	1.375
200	10.734	568×10^5	10.734

6600 Timing for Problem No. 3C

1. Case No. 2: (N = 300, 400, 500)

2. Notation:

$T_M = 100 * 10^{-9}$ sec $T_{MJ} = 1 * 10^{-6}$ sec $T_{MT} = \begin{cases} 100 * T_{MJ} \text{ average} \\ 200 * T_{MJ} \text{ maximum} \end{cases}$ $T_{EC} = \left[1.6 + 3.2\right] * 10^{-6}$ sec	T_{CP} = Control processor time I = Number of instructions executed T_{ECT} = Extended core transfer time with access $B = \begin{cases} 200 \text{ when } N = 300 \\ 150 \text{ when } N = 400 \\ 100 \text{ when } N = 500 \end{cases}$

3. Formulas:

$$T_{CP} = T_M * \left[136 + (N-1)\left(51 + \left(\frac{N+1}{2}\right)\left(43 + \left(\frac{N+1}{2}\right)\left(\frac{23}{2}\right) + N\left(\frac{23}{2}\right)\right)\right)\right.$$
$$\left. + N\left(73 + (N-1)\left(38 + \left(\frac{N+1}{2}\right)\left(\frac{18}{2}\right)\right)\right)\right]$$

$$I = \left[52 + (N-1)\left(17 + \left(\frac{N+1}{2}\right)\left(13 + \left(\frac{N+1}{2}\right)\left(\frac{12}{2}\right) + N\left(\frac{12}{2}\right)\right)\right)\right.$$
$$\left. + N\left(22 + (N-1)\left(12 + \left(\frac{N+1}{2}\right)\left(\frac{10}{2}\right)\right)\right)\right]$$

$$T_{ECT} = 4 * \left[T_{EC} + \left(T_M * N * B\right)\right]\left(\frac{N-B}{B}\right)\left(\frac{N}{B}\right)$$

$$T = T_{CP} + T_{ECT} + T_{MT}$$

4. Results:

N	T_{CP} (sec)	I	T_{ECT} (sec)	T (sec)
300	35.993	190×10^6	0.0180	36.011
400	84.995	451×10^6	0.1353	85.130
500	165.582	880×10^6	0.4008	165.982

6400 Timing for Problem No. 3C

1. Case No. 1: (N = 100, 200)

2. Notation:

$T_M = 100 * 10^{-9}$ sec $T_{MJ} = 1 * 10^{-6}$ sec $T_{MT} = \begin{cases} 100 * T_{MJ} \text{ average} \\ 200 * T_{MJ} \text{ maximum} \end{cases}$	T_{CP} = Central processor time I = Number of instructions executed T = Total job time

3. Formulas:

$$T_{CP} = T_M * \left[268 + (N-1) \left(92 + \left(\tfrac{N+1}{2}\right) \left(127 + \left(\tfrac{N+1}{2}\right) \left(\tfrac{180}{2}\right) + N \left(\tfrac{180}{2}\right) \right) \right) \right.$$
$$\left. + N \left(184 + (N-1) \left(145 + \left(\tfrac{N+1}{2}\right) \left(\tfrac{173}{2}\right) \right) \right) \right]$$

$$I = \left[52 + (N-1) \left(17 + \left(\tfrac{N+1}{2}\right) \left(13 + \left(\tfrac{N+1}{2}\right) \left(\tfrac{12}{2}\right) + N \left(\tfrac{12}{2}\right) \right) \right) \right.$$
$$\left. + N \left(22 + (N-1) \left(12 + \left(\tfrac{N+1}{2}\right) \left(\tfrac{10}{2}\right) \right) \right) \right]$$

$$T = T_{CP} + T_{MT}$$

4. Results:

N	T_{CP} (sec)	I	T (sec)
100	11.306	720×10^4	11.306
200	89.541	568×10^5	89.541

6400 Timing for Problem No. 3C

1. Case No. 2: (N = 300, 400, 500)

2. Notation:

$T_M = 100 * 10^{-9}$ sec $T_{MJ} = 1 * 10^{-6}$ sec $T_{MT} = \begin{cases} 100 * T_{MJ} \text{ average} \\ 200 * T_{MJ} \text{ maximum} \end{cases}$ $T_{EC} = [1.6 + 3.2] * 10^{-6}$ sec	T_{CP} = Control processor time T_{ECT} = Extended core transfer time with access T = Total job time $B = \begin{cases} 200 \text{ when } N = 300 \\ 150 \text{ when } N = 400 \\ 100 \text{ when } N = 500 \end{cases}$

3. Formulas:

$$T_{CP} = T_M * \left[268 + (N-1)\left(92 + \left(\tfrac{N+1}{2}\right)\left(127 + \left(\tfrac{N+1}{2}\right)\left(\tfrac{180}{2}\right) + N\left(\tfrac{180}{2}\right)\right)\right)\right.$$

$$\left. + N\left(184 + (N-1)\left(145 + \left(\tfrac{N+1}{2}\right)\left(\tfrac{173}{2}\right)\right)\right)\right]$$

$$I = \left[52 + (N-1)\left(17 + \left(\tfrac{N+1}{2}\right)\left(13 + \left(\tfrac{N+1}{2}\right)\left(\tfrac{12}{2}\right) + N\left(\tfrac{12}{2}\right)\right)\right)\right.$$

$$\left. + N\left(22 + (N-1)\left(12 + \left(\tfrac{N+1}{2}\right)\left(\tfrac{10}{2}\right)\right)\right)\right]$$

$$T_{ECT} = 4 * \left[T_{EC} + T_M * N * B\right]\left(\tfrac{N-B}{B}\right)\left(\tfrac{N}{B}\right)$$

$$T = T_{CP} + T_{ECT} + T_{MT}$$

4. Results:

N	T_{CP} (sec)	I	T_{ECT} (sec)	T (sec)
300	301.105	190×10^6	0.0180	301.123
400	712.497	451×10^6	0.1353	712.632
500	1390.151	880×10^6	0.4008	1390.552

Summary Problem 3C

6800 Timings

N	T_{CP} (sec)	I	T_{ECT} (sec)	T (sec)
100	0.344	720×10^4	—	0.344
200	2.683	568×10^5	—	2.683
300	8.998	190×10^6	0.0045	9.002
400	21.249	451×10^6	0.0383	21.278
500	41.396	880×10^6	0.1002	41.496

6600 Timings

N	T_{CP} (sec)	I	T_{ECT} (sec)	T (sec)
100	1.375	720×10^4	—	1.375
200	10.734	568×10^5	—	10.734
300	35.993	190×10^6	0.0180	36.011
400	84.995	451×10^6	0.1353	85.130
500	165.582	880×10^6	0.4008	165.982

6400 Timings

N	T_{CP} (sec)	I	T_{ECT} (sec)	T (sec)
100	11.306	720×10^4	—	11.306
200	89.541	568×10^5	—	89.541
300	301.105	190×10^6	0.0180	301.123
400	712.497	451×10^6	0.1353	712.632
500	1390.151	880×10^6	0.4008	1390.552

PROBLEM NO. 4

SOLUTION OF LINEAR EQUATIONS WITH TRI-DIAGONAL MATRIX

1. For $n = 200$, (200), 1000, obtain the time required to perform the following operations:

    a. $g_i = \dfrac{a_i}{c_i - b_i\, g_{i-1}}$, $\quad 2 \le i \le n-1$; $\quad g_1 = \dfrac{a_1}{c_1}$

    b. $h_i = \dfrac{b_i\, h_{i-1} - d_i}{c_i - b_i\, g_{i-1}}$, $\quad 2 \le i \le n$; $\quad h_1 = \dfrac{-d_1}{c_1}$

    c. $y_i = g_i\, y_{i+1} + h_i$, $\quad i = n-1,\ n-2,\ \cdots,\ 2,\ 1$; $\quad y_n = h_n$

2. It is assumed that a_i , b_i , c_i , d_i are given.

3. Illustrative Program:

```
        N1 = N - 1
        G(1) = A(1)/C(1)
        DO 1  I = 2, N1
        GG(I) = C(I) - B(I) * G(I-1)
    1   G(I) = A(I)/GG(I)
        GG(N) = C(N) - B(N) * G(N1)
        H(1) = - D(1)/C(1)
        DO 2  I = 2, N
    2   H(I) = ( B(I) * H(I-1) - D(I) )/GG(I)
        Y(N) = H(N)
        DO 3  I = 1, N1
        K = N - I
    3   Y(K) = G(K) * Y(K + 1) + H(K)
```

4. This problem is considered to be a subroutine or a function, not a job.

6800 Timing for Problem No. 4

1. Notation:

$T_M = 25 * 10^{-9}$ sec	T_{CP} = Central processor time
	I = Number of instructions executed

2. Formulas:

$$T_{CP} = T_M * [107 + (N-1)(40) + N(22)]$$

$$I = [39 + (N-1)(20) + N(11)]$$

3. Results

N	I	T_{CP} (ms)
200	6219	0.312
400	12419	0.622
600	18619	0.931
800	24819	1.242
1000	31019	1.551

6600 Timing for Problem No. 4

1. Notation

$T_M = 100 * 10^{-9}$ sec	T_{CP} = Central processor time
	I = Number of instructions executed

2. Formulas

$$T_{CP} = T_M + [107 + (N-1)(40) + N(22)]$$
$$I = [39 + (N-1)(20) + N(11)]$$

3. Results

N	I	T_{CP} (ms)
200	6219	1.247
400	12419	2.487
600	18619	3.727
800	24819	4.967
1000	31019	6.207

6400 Timing for Problem No. 4

1. Notation

$T_M = 100 * 10^{-9}$ sec	T_{CP} = Central processor time
	I = Number of instructions executed

2. Formulas

$$T_{CP} = T_M * [364 + (N-1)(277) + (N)(169)]$$
$$I = [39 + (N-1)(20) + (N)(11)]$$

3. Results

N	I	T_{CP} (ms)
200	6219	8.929
400	12419	17.849
600	18619	26.769
800	24819	35.689
1000	31019	44.609

EVALUATION OF POLYNOMIALS

1. Method:

   CASE 1: $N \leq 4$,   In line code.

   CASE 2:   N odd: but $N > 4$

   $$Y = \left(\left(a_n x + a_{n-1}\right) x^2 + a_{n-2} x + a_{n-3}\right) x^2 + \cdots a_o$$

   CASE 3:   N even: but $N > 5$

   $$Y = Y_o + Y_{\ell}$$

where $Y_o = \left(\left(a_{n-1} x + a_{n-3}\right) x^2 + a_{n-5}\right) x^2 + \cdots + a_{n-1} x$

   $Y_{\ell} = \left(\left(a_n x^2 + a_{n-2}\right) x^2 + a_{n-4}\right) x^2 + \cdots + a_o$

2. The evaluation of a polynomial is not considered to be a job but rather a subroutine or function.

6800 Timing for Problem No. 5

1. ### Notation

$T_M = 25 * 10^{-9}$ sec	T_{CP} = Central processor time
	I = Number of instructions executed

2. ### Formulas

$N = 2$

$$T_{CP} = T_M * 39$$
$$I = 8$$

$N = 5, 11$

$$T_{CP} = T_M * \left[60 + \left(\frac{N-3}{2} \right) (16) \right]$$
$$I = \left[19 + \left(\frac{N-3}{2} \right) (8) \right]$$

$N = 8$

$$T_{CP} = T_M * \left[75 + \left(\frac{N-4}{2} \right) (16) \right]$$
$$I = \left[23 + \left(\frac{N-4}{2} \right) (8) \right]$$

3. ### Results

N	I	T (μs)
2	8	0.975
5	27	1.900
8	39	2.675
11	51	3.100

6600 Timing for Problem No. 5

1. Notation

$T_M = 100 * 10^{-9}$ sec	T_{CP} = Central processor time
	I = Number of instructions executed

2. Formulas

$$N = 2 \qquad T_{CP} = T_M * 39$$

$$I = 8$$

$$N = 5, 11 \qquad T_{CP} = T_M * \left[60 + \left(\frac{N-3}{2}\right)(16)\right]$$

$$I = \left[19 + \left(\frac{N-3}{2}\right)(8)\right]$$

$$N = 8 \qquad T_{CP} = T_M * \left[75 + \left(\frac{N-4}{2}\right)(16)\right]$$

$$I = \left[23 + \left(\frac{N-4}{2}\right)(8)\right]$$

3. Results

N	I	T (μs)
2	8	3.900
5	27	7.600
8	39	10.700
11	51	12.400

1. Notation:

$T_M = 100 * 10^{-9}$ sec	T_{CP} = Central processor time I = Number of instructions executed

2. Formulas:

$N = 2$ $T_{CP} = T_M * 154$

            $I = 8$

$N = 5, 11$ $T_{CP} = T_M * \left[332 + \left(\dfrac{N-3}{2}\right)(160)\right]$

            $I = \left[19 + \left(\dfrac{N-3}{2}\right)(8)\right]$

$N = 8$ $T_{CP} = T_M * \left[404 + \left(\dfrac{N-4}{2}\right)(160)\right]$

            $I = \left[23 + \left(\dfrac{N-4}{2}\right)(8)\right]$

3. Results:

N	I	T (μs)
2	8	15. 400
5	27	49. 200
8	39	72. 400
11	51	97. 200

PROBLEM NO. 6
TABLE LOOK-UP PROCEDURE

1. Given a set of points X_i (abscissa stored monotonically), determine the pair of points X_L, X_R bounding the arguement T such that $X_L \le T \le X_R$ and $R = L + 1$.

2. Illustrative Example:

```
        IL = 1
        IR = N
1       M = (IL - IR)/2
        IF (T - X(M)) 2, 2, 3
2       IR = M
        IF (IR - IL - 1) 1, 4, 1
3       IL = M
        IF (IR - IL - 1) 1, 4, 1
4       RETURN
```

3. Find the time to determine the quantity IR in the above procedure, assuming that 10 searches are made.

4. A binary search was not used.

5. This routine was not considered to be a job but rather a subroutine or function.

6800 Timing for Problem No. 6

1. Notation:

$T_M = 25 * 10^{-9}$ sec	T_{CP} = Central processor time
K = 10	I = Number of instructions executed

2. Formulas:

$$T_{CP} = T_M * [53 + K (24)]$$
$$I = [16 + K (7)]$$

3. Results:

K	I	T_{CP} (µs)
10	86	7.325

6600 Timing for Problem No. 6

1. Notation

$T_M = 100 * 10^{-9}$ sec K = 10	T_{CP} = Central processor time I = Number of instructions executed

2. Formulas

$$T_{CP} = T_M * [53 + K (24)]$$
$$I = [16 + K (7)]$$

3. Results

K	I	T_{CP} (µs)
10	86	29. 3

6400 Timing for Problem No. 6

1. **Notation**

$T_M = 100 * 10^{-9}$ sec $K = 10$	T_{CP} = Central processor time I = Number of instructions executed

2. **Formulas**

$$T_{CP} = T_M * [105 + K (57)]$$
$$I = [16 + K (7)]$$

3. **Results**

K	I	T_{CP} (µs)
10	86	67. 5

83

1. For the problem below (item 2) show all subroutine linkages, prologues, etc. The timing for the problem should be from the beginning of Statement 19 to the beginning of Statement 21.

2. Problem:

```
1        FORMAT ( 3HOI = I4)
2        FORMAT ( 1X2F13. 5, E16. 7)
         -
         -
         -
19       DO 20  I = 1, 1000
         Y(I) = 5. + . 7 * SIN (2. * X(I) - 1. 57079)
         CALL POWER (Z, Y, X, I)
         WRITE (6, 1) I
20       WRITE (6, 2) X(I), Y(I), Z
21       GO TO 7
         -
         -
         -
         END
         -
         -
         -
         SUBROUTINE POWER (A, B, C, J)
         A = (B(J) * C(J) - 8. 5)/COS (C(J))
         RETURN
         END
```

3. All linkages and above statements where timed. See Appendix for listing of present code generation. The actual SINF and the WRITE STATEMENT timings are not included; however, all their linkages are. For prologues, refer to Volume III of this proposal containing documentation.

Timing for Problem No. 7A

1. Notation:

$T_M = \begin{cases} 25 * 10^{-9} \text{ sec, for } 6800 \\ 100 * 10^{-9} \text{ sec, for } 6600 \\ 100 * 10^{-9} \text{ sec, for } 6400 \end{cases}$	T_{CP} = Central processor time I = Number of instructions executed

2. Formulas:

$$T_{CP} = T_M * (295) * 10^4, \text{ for } 6800, 6600$$
$$T_{CP} = T_M * (620) * 10^4, \text{ for } 6400$$
$$I = (72) * 10^4$$

3. Results:

	6800	6600	6400
T_{CP} (ms)	64.8	259.0	620.0
I	720,000	720,000	720,000

1. For the problem below (item 2) show all subroutine linkages, prologues, etc. The timing for the problem should be from the beginning of Statement 49 to the beginning of Statement 51.

2. Problem:

```
            -
            -
            -
49          DO 50  I = 1, 100
            J = Z(I)/3.45
            X = FLOAT(2 * I - J) - 15.3
            Y = X/2.1 - FLOAT(MOD(I, J))

50          Z(I) = Z(I) + X * Y + X/Y - 20.3
51          GO TO 8
            -
            -
            -

            END
```

3. All linkages and code of the above statement where timed. See Appendix for listing of present code generation. Since the FLOAT function is less than 1 central memory word long, it is generated in line. For prologues refer to Volume III.

<u>Timing for Problem No. 7B</u>

1. <u>Notation</u>:

$T_M = \begin{cases} 25 * 10^{-9} \text{ sec, for } 6800 \\ 100 * 10^{-9} \text{sec, for } 6600 \\ 100 * 10^{-9} \text{sec, for } 6400 \end{cases}$	T_{CP} = Central processor time I = Number of instructions executed

2. <u>Formulas</u>:

$$T_{CP} = T_M * (284) * 100 \quad \text{for } 6800, 6600$$
$$T_{CP} = T_M * (759) * 100 \quad \text{for } 6400$$
$$I = (61) * 100$$

3. <u>Results</u>:

	6800	6600	6400
T_{CP} (ms)	0.71	2.84	7.59
I	6100	6100	6100

PROBLEM NO. 8

I. TAPE TO TAPE

1. a. Copy the information from one tape onto another tape.

    b. Number of characters on input tape = 20, 000, 000.

    c. Tape block size = 5, 000 characters.

2. Required Equipment:

    a. 607 tape units (150 inches/sec, 800 bpi).

    b. 1 peripheral and control processor.

    c. Data channels.

3. At this block size (5, 000 characters), 4000 records will fit onto 1 tape, hence 1 tape is required to hold all data.

4. The following cases have been timed:

    a. 1 tape input, 1 tape output

    b. 3 tapes input, 3 tapes output
        (2 reels with 1333 blocks, and 1 reel with 1334 blocks).

    c. 6 tapes input, 6 tapes output
        (4 reels with 667 blocks and 2 reels with 666 blocks).

Timing of Problem No. 8

1. Consider a tape, then

    a. between frames 16 µs is the periodicity at which a character has to be accepted.

    b. between blocks (INTER-RECORD GAP) there are 5 ms available to do the necessary bookkeeping such as testing for parity error, end of tape, end of file, etc.

2. Timing:

CASE 1: (1 tape) $1 * \left[\frac{5,000}{120,000} + \frac{3}{4 * 150} \right] * 4000$

CASE 2: (3 tapes) $3 * \left[\frac{5,000}{120,000} + \frac{3}{4 * 150} \right] * 1334$

CASE 3: (6 tapes) $6 * \left[\frac{5,000}{120,000} + \frac{3}{4 * 150} \right] * 667$

3. Timing Summary:

Number of Tapes Used	PP$_{6800}$ (sec)	PP$_{6600}$ (sec)	PP$_{6400}$ (sec)
1 tape	186.68	186.68	186.68
3 tapes	62.26	62.26	62.26
6 tapes	31.13	Not feasible*	Not feasible*

* The following timing considerations have to be made:

1. Time between 2 frames: 16.66 µs.

2. It takes two major cycles to issue an input or an output command.

3. It takes three major cycles to update a counter and transfer back to the original input command as stated in (1).

89

II.   TAPE TO MASS STORAGE TO PRINTER

1. a.   Read the tape on to a mass storage device.   Print the data from the mass storage device.

    b.   Print 2000 lines of 120 characters each.

    c.   Data is to be transferred from tape to mass storage in units of 1920 characters.

    d.   Data is to be accessed from mass storage for printing in units of 960 characters.

    e.   As much availability as desirable may be assumed for the mass storage device.

2. Assumptions:

    a.   There are 12 records of 1920 characters each on tape, the 13th record contains 960 characters, a total of 24,000 characters or 2,000 print lines of 120 characters each.

    b.   Find total time required using procedure outlined in step 1.

3. Procedure:

    a.   Read 1 block from tape: $\left[\frac{1.92}{120}\right]$ = 1.6 ms + 100 * T_{MJ} .

    b.   Transmit data read from tape into extended core storage: Steps required:

        i.   PP to CORE STORAGE: $[20 * 5 + 16]$ * T_{MJ}.

        ii.   CORE STORAGE to EXTENDED CORE: $\left[T_{EC} + T_M * 20\right] + T_{EJ} + T_{MT}$

    c.   Transmit data from extended core storage to core storage: $\left[T_{EC} + T_M * 10\right]$.

    d.   Transmit data from core storage to printer PP (which could be the same as the one that read the tape): $2 * [10 * 5 + 50] * T_{MJ}$

    e.   Print 1 line of data and continue steps a. and b. : $[50] * T_{MJ}$

4. What is the total time required to complete this job?

90

5. Requirements:

    1 Tape
    1 Printer
    2 Peripheral and control processors
    2 Data channels

6. Timing:

    a. Notation:

6800	6600/6400
$T_M = 25 * 10^{-9}$ sec	$T_M = 100 * 10^{-9}$ sec
$T_{MJ} = 250 * 10^{-9}$ sec	$T_{MJ} = 1 * 10^{-6}$ sec
$T_{EC} = [0.8 + 1.6] * 10^{-6}$ sec	$T_{EC} = [1.6 + 3.2] * 10^{-6}$ sec
$T_{EJ} = 2 * T_{MJ}$	$T_{EJ} = 2 * T_{MJ}$
$T_{MT} = \begin{cases} 100 * T_{MJ} \text{ average} \\ 200 * T_{MJ} \text{ maximum} \end{cases}$	$T_{MT} = \begin{cases} 100 * T_{HJ} \text{ average} \\ 200 * T_{MJ} \text{ maximum} \end{cases}$

    b. Formulas $\begin{cases} T_1 = 1 \text{ printer job time} \\ T_8 = 8 \text{ printers job time} \end{cases}$

$$T_{START} = \left[\frac{1,920}{120,000}\right] + 100 * T_{MJ} + \left[T_{EC} + T_M * 20\right] + \left[T_{EC} + T_M * 10\right]$$

$$+ T_{EJ} + T_{MT} + 2 * [10 * 5 + 50] * T_{MJ} + 50 * T_{MJ}$$

$$T_{CP} = \left[T_{EC} + T_M * 20\right] * 13 + \left[T_{EC} + T_M * 10\right] 25$$

$$T_{FIN} = \left[T_{EC} + T_M * 10\right] + 2 * [10 * 5 + 50] * T_{MJ} + 50 * T_{MJ}$$

$$T_1 = T_{PRINT} + T_{START} + T_{FIN}$$

$$T_{PRINT} = 2 \text{ sec at 2000 lines/min}$$

$$T_{READ} = T_{START} - \left[\frac{1,920}{120,000}\right]$$

$$T_8 = \frac{1}{8} T_{PRINT} + 7 * 50 * T_{MJ} + T_{START} + T_{FIN}$$

Time left to get back to tape read: $\left(5 \text{ ms} - T_{READ}\right)$.

c. Results:

	T_{START} (ms)	T_{CP} (µs)	T_{FIN} (µs)	T_{READ} (µs)	T_{PRINT} (sec)	T_1 (sec)	T_8
6800	16.119	105.2	65.2	119.0	2.0	120.16	15.02
6600	16.465	258.4	256.3	465.0	2.0	120.16	15.03
6400	16.465	258.4	256.3	465.0	2.0	120.16	15.03

III. CARD TO TAPE AND CARD

1. a. Four out of every five cards read into the processor are to be written on tape. The fifth card read in is to be punched as a duplicate.

   b. 2000 cards are to be input.

   c. Each input card has alpha-numeric information in all 80 columns of the card.

   d. Data from only one card is to be written in a single tape operation.

   e. Data recorded on tape is to be in a form which can subsequently be read in and printed without conversion. If main processor time is necessary to convert data from cards for printing, the time to accomplish this for the 1600 cards involved is to be added to the processor time. A statement as to whether or not the time for such conversion was included, is to be submitted.

2. Requirements:

   1 Tape
   1 Card reader
   1 Card punch
   1 Peripheral and control processor
   3 Data channels

3. Procedure:

   a. Read a card from the card reader. At 1200 cards/min. it takes 50 ms to read a card, or 600 μs/column. To convert 1 column takes 180 * T_{MJ} or at most 180 μs; this is accomplished between columns. The converted card is then output onto tape. The fifth card is read, converted and put into the punch buffer. After the card is converted it is punched.

   b. Total time required to perform this job is the card input time plus the punching of the last card:

   T = 50 * 2000 + 240   ms

   T = 100.24 sec

4. Results:

PP	T (sec)
6800	100.24
6600	100.24
6400	100.24

IV. REMOTE KEYBOARD-PRINTER

1. a. Twenty remote stations, each with a keyboard and printer, are to be simultaneously communicating with the processor in a full-duplex mode. Each message entered from the keyboard causes a message of equal size to be accessed from a random storage area in main memory. The message from mass memory is transmitted to the remote printer co-located with the given keyboard.

   b. One-half of the input messages are 50 characters and half are 100 characters. Sequence in which a given size record will occur is not known.

   c. The input buffer size allowed for each remote station cannot exceed 50 characters. Thus for the 100-character records, the same buffer is to be used for the first 50 and the last 50 characters.

   d. Each remote station is to transmit and receive 20 messages. No delay occurs between the input of one message and the next from a given station.

   e. If main processor time is consumed in multiplexing input and output, checking for each of messages, checking for full buffer area, etc., this time is to be added to the processor time.

2. Requirements:

   2 Peripheral and control processors
   2 Data channels
   Typewriter speed - 15 char/sec = 75 ms/char

3. Procedure:

   a. The peripheral and control processor polls each station for message ready, keeping back of the remote station polled:

   $(51) * T_{MJ}$ $(T_{MJ} = $ major cycle time) .

   b. Poll Print Message Buffer: $20 (13) * T_{MJ}$

   c. On Message Ready (Remote):

   i. Input message: $(25) * T_{MJ}$

   ii. Test for end of message: $(5) * T_{MJ}$
   NOT END: Resume with step a.

   iii. End of Message procedure: put message in proper buffer of central memory

   $= (5 * 5 + 1 * 5 + 32) * T_{MJ}$ if 50 char

   $= (10 * 5 + 1 * 5 + 32) * T_{MJ}$ if 100 char

94

iv. Access message of equal size from extended core:

$$= T_{MT} + T_{EJ} + \left[T_{EC} + T_M * 5 \right] \quad \text{if 50 char}$$

$$= T_{MT} + T_{EJ} + \left[T_{EC} + T_M * 10 \right] \quad \text{if 100 char}$$

$$T_{MT} = \begin{cases} 100 * T_{MJ} \text{ average} \\ 200 * T_{MJ} \text{ maximum} \end{cases}$$

$$T_{EJ} = 2 * T_{MJ}$$

$$T_{EC} = [0.8 + 1.6] * 10^{-6} \quad \text{if 6800,} \quad T_M = 25 * 10^{-9} \text{ sec}$$

$$= [1.6 + 3.2] * 10^{-6} \quad \text{if 6600/6400,} \quad T_M = 100 * 10^{-9} \text{ sec}$$

v. Set flag in print buffer message ready.

vi. Return to step a.

d. On MESSAGE READY in print buffer:

i. Transmit message to PP: $\quad [5 * 5 + 32] * T_M \quad$ if 50 char

$\qquad\qquad\qquad\qquad\qquad\quad [10 * 5 + 32] * T_M \quad$ if 100 char

ii. Print:

NOTE: Procedures a. and c. are done by the PP and procedures b. and d. are done by the other PP. In this manner both generations occur concurrently. Also total loop time for the remote PP:

$$T_{PP(REMOTE)_{max}} = 20 * (51) * T_{MJ} + 25 * T_{MJ} + 5 * T_{MJ}$$

$$+ 62 * T_{MJ} + 200\, T_{MJ} + 2\, T_{EJ}$$

$$T_{PP(R)_{max}} \qquad = 1314\, T_{MJ}$$

$$T_{PP(PRINT)} \qquad = \text{Printing} + 20\,(13) * T_{MJ}$$

$$T_{PP(PR)} \qquad = \text{Printing} + 260 * T_{MJ}$$

$$T_{CP} \qquad\qquad = 50 * \left(2 * T_{EC} + T_M * 5 + T_M * 10 \right)$$

Using this procedure, the remote PP could monitor up to 50 remote stations, and the printer PP could keep 30 printers busy.

4. Results:

	6800	6600	6400
T_{CP} (μs)	42.3	159.6	159.6

3.0 DEFINITION OF TERMS

B . . . Blocking factor

CM . . Central memory

CP . . Central processor

I . . . Number of instructions executed

I_o . . . Set-up of I/O request

ms . . Millisecond (10^{-3} sec)

ns . . . Nanosecond (10^{-9} sec)

P . . . Number of iterations

PP . . Peripheral and control processor

P_o . . Number of parameters in I/O request list (maximum = 6)

P_M . . Peripheral and control processor location

sec . . Second

T . . . Total job time

T_{EC} . . Extended core access and issue time for read/write

T_{ECT} . Extended core transfer time

T_{CP} . . Central processor time

T_D . . Disk transfer time (average), per 60-bit word

T_{DL} . . Disk latency time

T_{DT} . . Total disk time, including positioning, latency and transfer

T_{DP} . . Disk positioning time

T_{EJ} . . Time required to perform an exchange jump

$T_{I/O}$ . . Total I/O linkage and wait time

T_{ITRK} . Time required to perform K iterations

T_M . . Minor cycle time

T_{MJ} . . Major cycle time

T_{MT} . . Time required to terminate a job by monitor

T_{OV} . . Overlap time

W_o . . Wait time in CP, wait on I/O

μs . . . microsecond (10^{-6} sec)

1 byte = 6 bits

1 CM word = 10 bytes = 60 bits

1 PP word = 2 bytes = 12 bits

1 CM word = 5 PP words

4.0 APPENDIX

The following section contains several kernel problem listings, source and object codings and compilations generated on the 6600 computer system at Control Data's Los Angeles, California, facility.

The representative problems are as follows:

(1). . .   1A — Tiros Cloud Cover Analysis

(2). . .   1B — Tiros Picture Analysis

(3). . .   2A — Solution of Elliptic Partial Differential Equations (Variable Coefficients)

(4). . .   3A — Solution of Matrix Multiplication

(5). . .   4 — Solution of Linear Equations with Tri-diagonal Matrix

(6). . .   5 — Evaluation of Polynomials

(7). . .   7A — Fortran Problem A

(8). . .   7B — Fortran Problem B

PROBLEM NO. 1A

```
          ASCENT NASA1A
NASA1A    NO
START     SA1   MASK
          BX0   X1
          SB1   1
          SB7   480
          SB2   R0+R0
          SA1   R0+PS
          SA2   A1+B7
          SA3   A2+B7
          SA4   A3+B7
          SA6   R0+P2R-1
LOOP1A    BX5   X1*X0
          AX1   6
          SA4   A4+B2
          BX1   X1*X0
          SB2   B2+B1
          IX6   X1+X5
          BX5   X2*X0
          AX2   6
          SA1   A1+B2
          IX6   X6+X5
          BX2   X2*X0
          IX6   X6+X2
          BX5   X3*X0
          AX3   6
          SA2   A2+B2
          IX6   X6+X5
          BX3   X3*X0
          IX6   X6+X3
          SA3   A3+B2
          IX6   X6+X4
          SA6   A6+B2
          NE    B2 B7 LOOP1A
          SB5   5
          SA1   A6-479
          SB2   R0+R0
          SA0   A6-479
LOOP1B    BX6   X1
          LX1   12
          IX6   X6+X1
          SB2   B2+B1
          LX1   12
          IX6   X6+X1
          LX1   12
          IX6   X6+X1
          LX1   12
          IX6   X6+X1
          AX4   B5,X6
          BX7   X4*X0
          LX5   R0 X7
          SA1   A1+B2
          LX7   6
          BX6   X7+X5
          SA6   A0+B2
          NE    B2 B7 LOOP1B
          SA5   B0+P2R
          SA0   R0+P
          SA1   B0+A0
          SB6   12
          SB2   R0+R0
          SB3   R0+R0
```

99

```
              SB5   960
              SB4   B0+R0
LOOP2         SA2   A1+B7
              SA3   A1+B5
              BX6   X1-X5
              SA4   A2+B5
              SA1   A3+B5
              BX7   X2-X5
              SB2   B2+B1
              SA6   A2-B2
              SA7   A2
              BX6   X3-X5
              SA6   A3
              BX7   X4-X5
              SA7   A4
              NE    B2 B6 LOOP2
              SB3   B3+B1
              SB2   B0+R0
              SA0   A0+B3
              SA1   A0
              NE    B3 B7 LOOP2
              SB5   240
              SB6   47
              SA1   B0+P
              SA2   B0+P+1
              SB2   R0+R0
              SA0   B0+P-1
              SB3   2
              SB4   B0+R0
LOOP3         BX6   X1*X0
              AX1   6
              SB2   B2+B1
              BX5   X2*X0
              IX6   X6+X5
              AX2   6
              BX7   X1*X0
              SA1   A1+B3
              IX6   X7+X5
              BX4   X2*X0
              SA2   A1+B1
              IX6   X4+X6
              AX3   X6 B3
              BX7   X3*X0
              SA7   A0+B2
              NO
              NE    B2 B5 LOOP3
              SB4   B4+B1
              SB2   B0+R0
              SA0   A0+B7
              NE    B4 B6 LOOP3
              PS
MASK          CON   0077007700770077007700778
P             BSS   22560
PS            BSS   1921
P2B           BSS   480
              END
                                      ASCENT NASA1A
000000   46000                 NASA1A NO
000001   5110000046            START  SA1  MASK
              10011                    BX0  X1
000002   6110000001                    SB1  1
              6170000740                SB7  480
```

100

Address	Octal	Label	Instruction
000003	66200		SB2 B0+B0
	5110054107		SA1 B0+PS
	54217		SA2 A1+B7
000004	54327		SA3 A2+B7
	54437		SA4 A3+B7
	5160057707		SA6 B0+P2B-1
000005	11510	LOOP1A	BX5 X1*X0
	21106		AX1 6
	54442		SA4 A4+B2
	11110		BX1 X1*X0
000006	66221		SB2 B2+B1
	36615		IX6 X1+X5
	11520		BX5 X2*X0
	21206		AX2 6
000007	54112		SA1 A1+B2
	36665		IX6 X6+X5
	11220		BX2 X2*X0
	36662		IX6 X6+X2
000010	11530		BX5 X3*X0
	21306		AX3 6
	54222		SA2 A2+B2
	36665		IX6 X6+X5
000011	11330		BX3 X3*X0
	36663		IX6 X6+X3
	54332		SA3 A3+B2
	36664		IX6 X6+X4
000012	54662		SA6 A6+B2
	0527000005		NE B2 B7 LOOP1A
000013	6150000005		SB5 5
	5016777040		SA1 A6-479
000014	66200		SB2 B0+B0
	5006777040		SA0 A6-479
000015	10611	LOOP1B	BX6 X1
	20114		LX1 12
	36661		IX6 X6+X1
	66221		SB2 B2+B1
000016	20114		LX1 12
	36661		IX6 X6+X1
	20114		LX1 12
	36661		IX6 X6+X1
000017	20114		LX1 12
	36661		IX6 X6+X1
	23456		AX4 B5,X6
	11740		BX7 X4*X0
000020	22507		LX5 B0 X7
	54112		SA1 A1+B2
	20706		LX7 6
	12675		BX6 X7+X5
000021	54602		SA6 A0+B2
	0527000015		NE B2 B7 LOOP1B
000022	5150057710		SA5 B0+P2B
	5100000047		SA0 B0+P
000023	54100		SA1 B0+A0
	6160000014		SB6 12
	66200		SB2 B0+B0
000024	66300		SB3 B0+B0
	6150001700		SB5 960
	66400		SB4 B0+B0
000025	54217	LOOP2	SA2 A1+B7
	54315		SA3 A1+B5
	13615		BX6 X1-X5
	54425		SA4 A2+B5

```
000026  54135                               SA1   A3+B5
        13725                               BX7   X2-X5
              66221                         SB2   B2+B1
                   55622                    SA6   A2-B2
000027  54720                               SA7   A2
        13635                               BX6   X3-X5
              54630                         SA6   A3
                   13745                    BX7   X4-X5
000030  54740                               SA7   A4
        0526000025                          NE    B2 B6 LOOP2
              66331                         SB3   B3+B1
000031  66200                               SB2   B0+B0
        54003                               SA0   A0+B3
              54100                         SA1   A0
000032  0537000025                          NE    B3 B7 LOOP2
              6150000360                    SB5   240
000033  6160000057                          SB6   47
              5110000047                    SA1   B0+P
000034  5120000050                          SA2   B0+P+1
              66200                         SB2   B0+B0
000035  5100000046                          SA0   B0+P-1
              6130000002                    SB3   2
000036  66400                               SB4   B0+B0
000037  11610                         LOOP3 BX6   X1*X0
        21106                               AX1   6
              66221                         SB2   B2+B1
                   11520                    BX5   X2*X0
000040  36665                               IX6   X6+X5
        21206                               AX2   6
              11710                         BX7   X1*X0
                   54113                    SA1   A1+B3
000041  36675                               IX6   X7+X5
        11420                               BX4   X2*X0
              54211                         SA2   A1+B1
                   36646                    IX6   X4+X6
000042  23336                               AX3   X6 B3
        11730                               BX7   X3*X0
              54702                         SA7   A0+B2
                   46000                    NO
000043  0525000037                          NE    B2 B5 LOOP3
              66441                         SB4   B4+B1
                   66200                    SB2   B0+B0
000044  54007                               SA0   A0+B7
        0546000037                          NE    B4 B6 LOOP3
000045  0000000000                          PS
000046  007700770077007700770077    MASK    CON   007700770077007700770077B
000047  00000000000000054040        P       BSS   22560
054107  000000000000003601          PS      BSS   1921
057710  000000000000000740          P2B     BSS   480
060650                                      END

ERRORS      000000
SYMBOLS     000015
ASCENT      060650
ASPER-PP    000000
ASPER-CM    000000
000000  N    NASA1A 000001  N    START
```

PROBLEM NO. 1B

```
          ASCENT   NASA1B
NASA1B    NO
START     SB2   30
          SB5   B0-1
          SB4   6
          SB1   1
          SX1   B1+B0
          AX6   60
          LX2   B2 X1
          SB3   B4+B4
          IX3   X1+X2
          LX4   B3 X3
          SA6   GREY
          IX5   X3+X4
          LX0   B4 X5
          IX2   X0+X5
          LX3   B3 X2
          SX7   65
          IX4   X3+X7
          SX1   8
TRAN      IX6   X4+X6
          SX1   X1+B5
          SA6   A6+B1
          IX7   X6+X4
          SA7   A6+B1
          IX6   X4+X6
          SA6 A7+B1
          IX7   X6+X4
          SA7   A6+B1
          NZ    X1 TRAN
          SA1   A6-31
          SA2   R
          SB7   10
          AX6   60
          SB5   B0+R0
          SA0   A2
          SX4   32
          BX3   X2-X1
          SA6   B0+S-1
          SB6   40
          SB3   100
          SB4   B0+R0
          MX5   54
INTA      SX4   X4-1
          SB2   B0+R0
INTB      BX0   -X5*X3
          AX3   6
          SB5   B5+1
          NZ    X0     NEXT
          SX6   X6+1
NEXT      NE    B7 B5 INTB
          SA2   A2+B1
          SB4   B4+B1
          BX3   X2-X1
          SB5   B0+R0
          NE    B6 B0 INTB
          SA6   A6+B1
          SB2   B1+B2
          SB4   B0+R0
          AX6   60
          NE    B3 B2 INTB
          SA1   A1+1
```

```
            SA2   A0+B0
            BX3   X2-X1
            NZ    X4     INTA
            SA1   A6-99
            SB5   32
            SX2   400
            SX3   B0+B1
            SB7   B0+B5
            SX4   B0+B3
INTC        SA1   A1-3099
            BX6   X2
            FX4   X4-X3
INTD        SB7   B7-B1
            SA1   A1+B3
            IX6   X6-X1
            NE    B7 B0 INTD
            SA6   A6+B1
            SB7   B0+B5
            ZR    X4     INTC
            PS
GREY        BSS   33
R           BSS   5761
S           BSS   3300
            END

                                       ASCENT  NASA1B
000000  46000                NASA1B    NO
000001  6120000036           START     SB2   30
                6150777776             SB5   B0-1
000002  6140000006                     SB4   6
                6110000001             SB1   1
000003  76110                          SX1   B1+B0
                21674                   AX6   60
                22221                   LX2   B2 X1
                    66344              SB3   B4+B4
000004  36312                          IX3   X1+X2
                22433                   LX4   B3 X3
                    5160000037         SA6   GREY
000005  36534                          IX5   X3+X4
                22045                   LX0   B4 X5
                    36205              IX2   X0+X5
                        22332          LX3   B3 X2
000006  7170000101                     SX7   65
                36437                   IX4   X3+X7
000007  7110000010                     SX1   8
000010  36646                TRAN      IX6   X4+X6
                73115                   SX1   X1+B5
                    54661              SA6   A6+B1
                        36764          IX7   X6+X4
000011  54761                          SA7   A6+B1
                36646                   IX6   X4+X6
                    54671              SA6   A7+B1
                        36764          IX7   X6+X4
000012  54761                          SA7   A6+B1
                0311000010             NZ    X1 TRAN
000013  5016777740                     SA1   A6-31
                5120000100             SA2   R
000014  6170000012                     SB7   10
                21674                   AX6   60
                    66500              SB5   B0+B0
000015  54020                          SA0   A2
                7140000040             SX4   32
                        13321          BX3   X2-X1
```

104

```
000016  5160013300                      SA6   B0+S-1
                6160000050              SB6   40
000017  6130000144                      SB3   100
                66400                   SB4   B0+B0
                    43566               MX5   54
000020  7244777776          INTA        SX4   X4-1
                66200                   SB2   B0+B0
000021  15035               INTB        BX0   -X5*X3
            21306                       AX3   6
                6155000001              SB5   B5+1
000022  0310000023                      NZ    X0        NEXT
                7266000001              SX6   X6+1
000023  0575000021          NEXT        NE    B7 B5  INTB
            54221                       SA2   A2+B1
                66441                   SB4   B4+B1
000024  13321                           BX3   X2-X1
            66500                       SB5   B0+B0
                0560000021              NE    B6 B0  INTB
000025  54661                           SA6   A6+B1
            66212                       SB2   B1+B2
                66400                   SB4   B0+B0
                    21674               AX6   60
000026  0532000021                      NE    B3 B2  INTB
                5011000001              SA1   A1+1
000027  54200                           SA2   A0+B0
            13321                       BX3   X2-X1
                0314000020              NZ    X4        INTA
000030  5016777634                      SA1   A6-99
                6150000040              SB5   32
000031  7120000620                      SX2   400
                76301                   SX3   B0+B1
                66705                   SB7   B0+B5
000032  76403                           SX4   B0+B3
000033  5011771744          INTC        SA1   A1-3099
            10622                       BX6   X2
                31443                   FX4   X4-X3
000034  67771               INTD        SB7   B7-B1
            54113                       SA1   A1+B3
                37661                   IX6   X6-X1
000035  0570000034                      NE    B7 B0  INTD
            54661                       SA6   A6+B1
                66705                   SB7   B0+B5
000036  0304000033                      ZR    X4        INTC
                0000000000              PS
000037  0000000000000000000041  GREY    BSS   33
000100  0000000000000013201     R       BSS   5761
013301  0000000000000006344     S       BSS   3300
021645                                  END
ERRORS      000000
SYMBOLS     000016
ASCENT      021645
ASPER-PP    000000
ASPER-CM    000000
000000  N   NASAIR 000001  N   START
```

PROBLEM NO. 2A

```
              ASCENT   NASA3A
NASA3A   NO
START    SX6  10000
         SA6  INTER
PSTART   SX7  3
         SA7  NZL
         SX6  B0+A0
         SA6  NZK
         SB1  B0+1
         SB2  B0+25
         SB3  B0+625
         SB7  B0+23
         SB6  B0+3125
         SA6  D
KLOOP    SA0  B0+U+651
         SA5  B0+W1+651
         SA1  A0+B1
         SA2  A5+B6
         SA3  A0-B1
         SA4  A2+B6
         SA6  TEMP-1
IJLOOP1  FX0  X5*X1
         FX7  X2*X3
         SA1  A0+B2
         SA2  A4+B6
         FX0  X0+X7
         SA3  A0-B2
         SA5  A2+B6
         FX0  X4*X1
         SA1  A5+B6
         FX7  X2*X3
         SA4  A0-B6
         SA2  A5+B6
         SA0  A0+B1
         SB4  B4+B1
         FX6  X6+X0
         FX0  X4*X5.
         SA5  A0+B6
         SA3  A0-B1
         FX6  X6+X7
         FX7  X2*X1
         SA2  A5+B6
         SA1  A0+B1
         FX6  X6+X0
         SA4  A2+B6
         FX6  X6+X7
         SA6  A6+B1
         NE   B4 B7 IJLOOP1
         SA6  A6+B1
         SB5  B5+B1
         SA1  A1+B1
         SA2  A2+B1
         SA3  A3+B1
         SA4  A4+B1
         SA5  A5+B1
         NE   B5 B7 IJLOOP1
         SA5  TEMP
         SA0  B0+U+651
         SA1  B0+W5+651
         SA2  A0+B6
         SA3  A1+B6
         NX5  X5,B0
```

```
            SA4     A0+B0
  IJLOOP2   FX0     X1*X2
            FX1     X5*X0
            FX7     X3*X4
            BX6     X4
            SA6     A4-625
            SB4     B4+B1
            SA0     A0+B1
            SA2     B0+D
            SA3     A3+B1
            FX6     X1+X7
            NX7     B0,X6
            SA7     A4+A0
            FX6     X6-X4
            AX0     X6,B3
            BX4     -X0-X6
            FX7     X4+X2
            SA7     B0+D
            SA1     A1+B1
            SA2     A0+B6
            SA4     A0+A0
            NE      B4 B7 IJLOOP2
            SB5     B5+B1
            SB4     B0+B0
            SA1     A1+B1
            SA2     A2+B1
            SA3     A3+B1
            SA4     A4+B1
            SA5     A5+B1
            NE      B5 B7 IJLOOP2
     +      SA1     NZL
            SX2     B0+B1
            IX0     X1-X2
            BX7     X0
            SB5     B0+B0
            SA7     B0+NZL
            NZ      X0 KLOOP
     -      SA1     B0+ITER
            SX2     B0+B1
            IX0     X1-X2
            BX6     X0
            SB5     B0+B0
            SA6     B0/ITER
            SX7     B0+3
            SA7     NZL
            NZ      X0 PSTART
            PS
  ITER      BSS     1
  NZL       BSS     1
  NZK       BSS     1
  D         BSS     1
  U         BSS     3125
  W1        BSS     3125
  W2        BSS     3125
  W3        BSS     3125
  W4        BSS     3125
  W6        BSS     3125
  W8        BSS     3125
  F         BSS     3125
  W5        BSS     3125
  W7        BSS     3126
  TEMP      BSS     625
```

```
PROB3A    SB7  80/100
          END
                              ASCENT  NASA3A
000010   46000          NASA3A    NO
000001   7160023420     START     SX6  10000
          5160076257              SA6  INTER
000002   7170000003     PSTART    SX7  3
          5170000047              SA7  NZL
000003   76600                    SX6  B0+B0
          5160000050              SA6  NZK
000004   6110000001               SR1  B0+1
          6120000031              SR2  B0+25
000005   6130001161               SR3  B0+625
          6170000027              SB7  B0+23
000006   6160006065               SB6  B0+3125
          5160000051              SA6  D
000007   5100001265     KLOOP     SA0  B0+U+651
          5150007352              SA5  B0+W1+651
000010   54101                    SA1  A0+B1
          54256                   SA2  A5+B6
          55301                   SA3  A0-B1
          54426                   SA4  A2+B6
000011   5160075074               SA6  TEMP-1
000012   40051          IJLOOP1   FX0  X5*X1
          40723                   FX7  X2*X3
          54102                   SA1  A0+B2
          54246                   SA2  A4+B6
000013   30007                    FX0  X0+X7
          55302                   SA3  A0-B2
          54526                   SA5  A2+B6
          40041                   FX0  X4*X1
000014   54156                    SA1  A5+B6
          40723                   FX7  X2*X3
          55406                   SA4  A0-B6
          54256                   SA2  A5+B6
000015   54001                    SA0  A0+B1
          66441                   SB4  B4+B1
          30660                   FX6  X6+X0
          40045                   FX0  X4*X5
000016   54506                    SA5  A0+B6
          55301                   SA3  A0-B1
          30667                   FX6  X6+X7
          40721                   FX7  X2*X1
000017   54256                    SA2  A5+B6
          54101                   SA1  A0+B1
          30660                   FX6  X6+X0
          54426                   SA4  A2+B6
000020   30667                    FX6  X6+X7
          54661                   SA6  A6+B1
          0547000012              NE   B4 B7 IJLOOP1
000021   54661                    SA6  A6+B1
          66551                   SB5  B5+B1
          54111                   SA1  A1+B1
          54221                   SA2  A2+B1
000022   54331                    SA3  A3+B1
          54441                   SA4  A4+B1
          54551                   SA5  A5+B1
000023   0557000012               NE   B5 B7 IJLOOP1
          5150075075              SA5  TEMP
000024   5100001265               SA0  B0+U+651
          5110062135              SA1  B0+W5+651
000025   54206                    SA2  A0+B6
```

```
                 54316                       SA3   A1+B6
                     24505                   NX5   X5,B0
                         54400               SA4   A0+B0
        000026  40012               IJLOOP2  FX0   X1*X2
                   30150                      FX1   X5+X0
                       40734                  FX7   X3*X4
                           10644              BX6   X4
        000027  5064776616                    SA6   A4-625
                     66441                    SB4   B4+B1
                         5400:                SA0   A0+B1
        000030  5120000051                    SA2   B0+D
                     54331                    SA3   A3+B1
                         3061;                FX6   X1+X7
        000031  24706                         NX7   B0,X6
                   54740                      SA7   A4+B0
                       31664                  FX6   X6-X4
                           2303(              AX0   X6,B3
        000032  17460                         BX4   -X0-X6
                   30742                       FX7   X4+X2
                       5170000051             SA7   B0+D
        000033  54111                         SA1   A1+B1
                   54206                      SA2   A0+B6
                       54400                  SA4   A0+B0
        000034  0547000026                    NE    B4 B7 IJLOOP2
                     66551                    SB5   B5+B1
                         66400                SB4   B0+B0
        000035  54111                         SA1   A1+B1
                   54221                      SA2   A2+B1
                       54331                  SA3   A3+B1
                           54441              SA4   A4+B1
        000036  54551                         SA5   A5+B1
                     0557000026               NE    B5 B7 IJLOOP2
        000037  5110000047           +        SA1   NZL
                   76201                      SX2   B0+B1
                       37012                  IX0   X1-X2
        000040  10700                         BX7   X0
                   66500                      SB5   B0+B0
                       5170000047             SA7   B0+NZL
        000041  0310000007                    NZ    X0 KLOOP
        000042  5110000046           +        SA1   A0+ITER
                   76201                      SX2   B0+B1
                       37012                  IX0   X1-X2
        000043  10600                         BX6   X0
                   66500                      SB5   B0+B0
K                    5160000046               SA6   B0/ITER
        000044  7170000003                    SX7   B0+3
                   5170000047                 SA7   NZL
        000045  0310000002                    NZ    X0 PSTART
                   0000000000                 PS
        000046  00000000000000000001 ITER     BSS   1
        000047  00000000000000000001 NZL      BSS   1
        000050  00000000000000000001 NZK      BSS   1
        000051  00000000000000000001 D        BSS   1
        000052  00000000000000006065 U        BSS   3125
        006137  00000000000000006065 W1       BSS   3125
        014224  00000000000000006065 W2       BSS   3125
        022311  00000000000000006065 W3       BSS   3125
        030376  00000000000000006065 W4       BSS   3125
        036463  00000000000000006065 W6       BSS   3125
        044550  00000000000000006065 W8       BSS   3125
        052635  00000000000000006065 F        BSS   3125
        060722  00000000000000006065 W5       BSS   3125
```

109

```
              067007   00000000000000006066      W7       BSS   3126
              075075   00000000000000001161      TEMP     BSS   625
              076256   6170000144                PROB3A   SR7   R0/100
                                                          END

ERRORS        000002
SYMBOLS       000032
ASCENT        076260
ASPER-PP      000000
ASPER-CM      000000
000000   N    NASA3A 000001   N      START 014224   N          W2 022311   N          W3
030376   N           W4 036463   N.         W6 044550   N          W8 052635   N          F
067007   N           W7 076256   N      PROB3A 076257   U    INTER
```

PROBLEM NO. 3A

```
5                       ASCENT  PROB3A
          START     SA6 B0+C-1
                    SA3   B0+A-100
                    SA4 B0+B-1
                    SB2 B0+B0
                    SB1 B0+1
                    SB3  B0/9999
                    SB5 B0+B0
                    SB4 B0+B0
                    SB6  B0+9000
          LOOP      SA1 A3+B7.A(I,K)
                    SA2 A4+B1.B(K,J)
                    FX7 X5+X7.ACCUMULATE TOTAL
                    FX6 X3*X4.A(I,K+1)*B(K+1,J)
                    SA3 A1+B7.A(I,K+1)
                    SB5 B5+B1.R=R+2
                    FX5 X0+X6.ADD  #*#S
                    SA4 A2+B1.B(K+1,J)
                    FX0 X1*X2.A(I,K)*B(K,J)
                    NE  B5B7 LOOP
                    FX7 X7+X6
                    NX6 B0,X7
                    SA6 A6+B1
                    SB4 B4+B1
                    SB5 B0+B0
                    SA4 A2+B1
                    SA3 A1-B6
                    NE   B4 B7 LOOP
                    SB2 B2+B7
                    SA3 A3+B1
                    SA4 A4-B3
                    SB4 B0+B0
                    NE  B2 B7 LOOP
                    PS
                    PS
                    PS
                    BSS  100
          C         BSS  10000
                    BSS  100
          A         BSS  10000
          B         BSS  10000
                    BSS  100
                    END
```

000000	5160000160	START	SA6 B0+C-1
	5130023601		SA3 B0+A-100
000001	5140047364		SA4 B0+B-1
	66200		SB2 B0+B0
000002	6110000001		SB1 B0+1
	6130023417		SB3 B0/9999
000003	66500		SB5 B0+B0
	66400		SB4 B0+B0
	6160021450		SB6 B0+9000
000004	54137	LOOP	SA1 A3+B7.A(I,K)
	54241		SA2 A4+B1.B(K,J)
	30757		FX7 X5+X7.ACCUMULATE TOTAL
	40634		FX6 X3*X4.A(I,K+1)*B(K+1,J)
000005	54317		SA3 A1+B7.A(I,K+1)
	66551		SB5 B5+B1.R=R+2
	30506		FX5 X0+X6.ADD #*#S
	54421		SA4 A2+B1.B(K+1,J)
000006	40012		FX0 X1*X2.A(I,K)*B(K,J)

ASCENT PROB3A

111

```
UF                   0500073151          NE    B5B7 LOOP
                            3n776        FX7   X7+X6
        000007  24607                    NX6   B0,X7
                54661        .           SA6   A6+B1
                       66n41             SB4   B4+B1
                             66500       SB5   B0+B0
        000010  54421                    SA4   A2+B1
                .       55316            SA3   A1-B6
                         0547000004      NE    B4 B7 LOOP
        000011  66227                    SB2   B2+B7
                54331.                   SA3   A3+B1
                       55443             SA4   A4-B3
                             66400       SB4   B0+B0
        000012  0527000n4                NE    B2 B7 LOOP
                       00n0000000        PS
        000013  0000000n0n               PS
        000014  0000000n0n               PS
        000015  0000000nn0n000n000144    BSS   100
        000161  0000000nn0000n023420  C  BSS   10000
        023601  0000000nn0000n000144     BSS   100
        023745  n000000nn0000n023420  A  BSS   10000
        047365  n000000nn0000n023420  B  BSS   10000
        073005  n0000000nn0000n000144    BSS   100
        073151·                          END
ERRORS         000001
SYMBOLS        000011
ASCENT         073152
ASPER=PP       000000
ASPER-CM       000000
000000  N      START 073151   U    B5B7
```

PROBLEM NO. 4

```
                SA7 A0+B7.+1(N) Y(N)
                SB2 B0+B0.I=N1
                SA1 B6+G .G(K)=X1
                SA2 B6+H .H(K)=X2
LOOP3           FX0 X1*X7.G(K)*Y(K+1)
                FX3 X0+X2.G(K)*Y(K+1)+H(K)
                SA1 A1-B1.NEXT G(K)-1
                SA2 A2-B2.NEXT H(K)-1
                BX7 X3    .SAVE PREVIOUS Y
                SB2 B2+B1.I=I+1
                NX6 B0,X3.NORMALIZE
                SA6 A7-B1.STORE
                NE  B2 B6 LOOP3  .LOOP
                PS
                PS
N               CON 200
                BSS 1
A               BSS 200
                BSS 1
C               BSS 200
                BSS 1
B               BSS 200
                BSS 1
G               BSS 200
                BSS 1
D               BSS 200
                BSS 1
H               BSS 200
                BSS 1
Y               BSS 200
                BSS 1
                END
```

```
                                           ASCENT  PROB4
000000  6140777776              PROB4      SB4 B0-1
        5150000034                         SA5 B0+A  .A(1)
000001  5130000345                         SA3 B0+C  .C(1)
        44653                              FX6 X5/X3.A(1)/C(1)
000002  5140000656                         SA4 B0+B  .B(1)
        46000                              NO        .PAN
000003  6110000001                         SB1 B0+1  .1
        6120000000                         SB2 B0+0  .I=0
000004  5101001167                         SA0 B1+G  .STORE ADDR FOR G
        73050                              SX0 B0+X5.X0=X5
000005  5110000032                         SA1 B0+N  .N
        63714                              SB7 X1+B4
        67671                              SB6 B7-B1.B6=N-2
000006  54600                              SA6 A0+B0.G(1)=A(1)/C(1)
 ··· ··· ··    46000                       NO
        46000                              NO
000007  44750                   LOOP1      FX7 X5/X0.A(I-1)/GG(I-1)
        40146                              FX1 X4*X6.B(I)*G(I-1)
        31231                              FX2 X3-X1.C(I)-B(I)*G(I-1)
        10077                              BX0 X7    .G(I) G(I-1)
·· 000010  54551                           SA5 A5+B1.A(I)
        54331                              SA3 A3+B1.*C(I+1)
        54441                              SA4 A4+B1.*B(I+1)
        66221                              SB2 B2+B1.I=I+1
000011  24602                              NX6 B0,X2.NORMALIZE
        54602                              SA6 A0+B2.STORE G(I-1)
        0526000007                         NE  B2 B6 LOOP1
000012  5150001500                         SA5 B0+D  .*D(1)
 ·· ·· ··    5120000345                    SA2 B0+C  .C(1)
```

114

```
000013  44052                    FX0  X5/X2.D(1)/C(1)=H(1)
        40146                    FX1  X4*X6.B(N)*G(N1)
                31431            FX4  X3-X1.C(N)-B(N)*G(N1)
                        24604    NX6  B0,X4.NORMALIZE
000014  54200                    SA2  B0+A0.G(-1)
        5131000656               SA3  B1+B .B(2)
                        54551    SA5  A5+B1.D(2)
000015  10722                    RX7  X2   .G(-1) X7
        54221                    SA2  A2+B1.*G(1)
                5110000656       SA1  B0+B .*B(1)
000016  10644                    RX6  X4   .X4 X6
        54720                    SA7  A2+B0.G(-1) G(1)
                54661            SA6  A6+B1.G(N)=G(NN)=CN-BN*G(N1)
                        14600    BX6  -X0  .-D(1)/C(1)
000017  5100002011               SA0  B0+H.*ADDR OF H
        66200                    SB2  B0+B0.I=0
                        55601    SA6  A0-B1. +1(-1)
000020  44702           LOOP2    FX7  X0/X2.(B(I)*H(I-1)-D(I)F
        40163                    FX1  X6*X3.B(I)*H(I-1) T
                31415            FX4  X1-X5.B(I)*+1(I-1)-D(I) T
                        24604    NX6  B0,X4.NORMALIZE X6
000021  10066                    BX0  X6   .X6 X0
        54551                    SA5  A5+B1.D(I)+1
                54331            SA3  A3+B1.B(I)+1
                        54221    SA2  A2+B1.G(I)+1
000022  66221                    SB2  B2+B1.I=I+1
        54702                    SA7  A0+B2. +1(2)
                0527000020       NE   R2 B7 LOOP2  .LOOP
000023  55101                    SA1  A0-B1.+1(-1)
        10611                    BX6  X1   . X6
                5100002322       SA0  B0+Y .SET Y(1) ADDR=A0
000024  54611                    SA6  A1+B1.+1(-1) +1(1)
        54707                    SA7  A0+B7.+1(N) Y(N)
                66200            SB2  B0+B0.I=N1
000025  5116001167               SA1  B6+G .G(K)=X1
        5126002011               SA2  B6+H .H(K)=X2
000026  40017           LOOP3    FX0  X1*X7.G(K)*Y(K+1)
        30302                    FX3  X0+X2.G(K)*Y(K+1)+H(K)
                55111            SA1  A1-B1.NEXT G(K)-1
                55222            SA2  A2-B2.NEXT H(K)-1
000027  10733                    RX7  X3   .SAVE PREVIOUS Y
        66221                    SB2  B2+B1.I=I+1
                24603            NX6  B0,X3.NORMALIZE
                        55671    SA6  A7-B1.STORE
000030  0526000026               NF   R2 B6 LOOP3  .LOOP
        0000000000               PS
000031  0000000000               PS
000032  00000000000000000000310  N    CON  200
000033  00000000000000000001          RSS  1
000034  00000000000000000000310  A    RSS  200
000344  00000000000000000001          RSS  1
000345  00000000000000000000310  C    RSS  200
000655  00000000000000000001          RSS  1
000656  00000000000000000000310  B    RSS  200
001166  00000000000000000001          RSS  1
001167  00000000000000000000310  G    RSS  200
001477  00000000000000000001          RSS  1
001500  00000000000000000000310  D    RSS  200
002010  00000000000000000001          RSS  1
002011  00000000000000000000310  H    RSS  200
002321  00000000000000000001          RSS  1
002322  00000000000000000000310  Y    RSS  200
```

```
       002632  00000000n00000000001          BSS  1
       002633                                 END
ERRORS         000000
SYMBOLS        000017
ASCENT         002633
ASPER-PP       000000
ASPER-CM       000000
000000   N     PROB4
```

PROBLEM NO. 5

```
5                     ASCENT   PROB5
                      SB1   B0+B0
                      SB5   B0+1000
                      SB7   B0+4
                      SB1   B0+1
                      SB4   B0+B0
                      SB2   B1+B1
                      SB6   B0+B0
          LOOP2       SA1   B0+Y
                      FX0   X1*X1
                      SA2   A1+B1
                      FX7   X1*X2
                      SA5   A1+B2
                      SA4   A2+B2
          LOOP1       FX6   X7+X5
                      FX7   X0*X6
                      FX2   X1*X4
                      SA3   A5+B2
                      SB6   B6+B2
                      FX5   X2+X3
                      SA4   A4+B2
                      NO
                      NE    B6  B7 | LOOP1
                      SB4   B4+B1
                      SB6   B0+B0
                      NE    B4  B5 LOOP2
                      PS
                      PS
                      BSS   1
           Y          BSS   100
                      BSS   1
                      END
                                        ASCENT   PROB5
000000   66100                          SB1   B0+B0
            6150001750                   SB5   B0+1000
000001   6170000004                      SB7   B0+4
              A110000001                 SB1   B0+1
000002   66400                           SB4   B0+B0
              66211                       SB2   B1+B1
                66600                      SB6   B0+B0
000003   5110000013          LOOP2       SA1   B0+Y
              40011                        FX0   X1*X1
                54211                       SA2   A1+B1
000004   40712                            FX7   X1*X2
              54512                         SA5   A1+B2
                54422                        SA4   A2+B2
000005   30675               LOOP1       FX6   X7+X5
              40706                        FX7   X0*X6
                40214                       FX2   X1*X4
                  54352                      SA3   A5+B2
000006   66662                            SB6   B6+B2
              30523                        FX5   X2+X3
                54442                       SA4   A4+B2
                  46000                      NO
000007   0567000005                      NE    B6  B7 LOOP1
              66441                        SB4   B4+B1
                66600                      SB6   B0+B0
000010   0545000003                      NE    B4  B5 LOOP2
              0000000000                   PS
000011   0000000000                       PS
000012   00000000000000000001            BSS   1
000013   00000000000000K0000144   Y      BSS   100
```

117

```
        000157  0000000n0n0n0n0000001          BSS  1
        000160                                  END
ERRORS          000000
SYMBOLS         000006
ASCENT          000160
ASPER-PP        000000
ASPER-CM        000000
```

PROBLEM NO. 7A

```
                FORTRAN 4 PROGRAM SEVENA (OUTPUT,TAPE 6=OUTPUT)
*NM
                PROGRAM SEVENA
*NM
                DIMENSION X(1000),Y(1000)
              1 FORMAT(3H0I=I4)
              2 FORMAT(1X2F13.5,E16.7)
                DO 10 I=1,1000

000000    7110000001   200002
          10610
          76700
000001    0400200004
000002    5110340003   200003
          7120000001
000003    7130001750
          36612
          37736
000004    5160340003   200004
          0337360001

          10 X(I)=1.2

000005    9140340003   200005
          6214320003
000006    5150260007  .
          10650
          56610
000007    0400200003

          19 DO 20 I=1,1000

000010    7110000001   200006
          10610
          76700
000011    0400200011
000012    5110340003   200010
          7120000001
000013    7130001750
          36612
          37736
000014    5160340003   200011
          0337360002

          Y(I)=5.+7.*SIN(2.*X(I)-1.57079)

000015    10460
          6214320003
000016    5150260012
          56110
000017    5120260013
          41651
          31662
000020    24606
          5160240002
000021    6110240002
          0100400100
000022    10160
          5130340003
000023    6213320004 .
          5140260010 .
000024    5150260011
```

119

```
              10640
              41751
   000025     30667
              24606
              56610

              CALL POWER(Z,Y,X,I)

   000026     6110340004
              6120320002
   000027     6130320001
             ‾6140340003
   000030     0100400200

                 WRITE (6,1),I
*UF

 ‾000031‾ 6130260001

          20     WRITE (6,2),X(I),Y(I),Z
*UF

   000032     6130260004   200015
             ‾0400200010

          21 GO TO 7

 ··000033   0400360003   200016

             7 STOP

   000034     5120260014   200017
              10720
 ‾000035     0100400300

                 END

   000036     5130260014
              10730
 ‾000037··  0100400400
```

```
                    SUBROUTINE POWER(A,B,C,J)

004003   0                  200002
004004   0                  200003
004005   0                  200004
004006   0                  200005
004007   0                  200006
004010   0                  200007
004011   0                  200001
004012   76710
         76120
         20122
         36717
004013   76230
         20244
         36727
004014   5170240001
         76740
004015   5170240002

                    DIMENSION B(100),C(100)
                    A=(B(J)*C(J)-A.5)/COS(C(J))

         56340
004016   6173777776
         63737
         76670
004017   5160220001
         5140220001
004020   63140
         0100400500
004021   5150240001    200011
         5110240002
004022   63150
         21522
         63250
         21522
004023   63350
         63410
         5160240003
004024   56240
         6152777776
         63525
004025   56350
         5140220001
         53440
004026   5150260001
         41634
         31665
004027   24606
         10160
         5120240003
004030   45612
         56610

                    RETURN

         0400200001

              J   END

004031   5130260002
```

121

```
        10730
 004032    0100400400


00.22.57. NASA7A . READ.
00.22.58. NASA7A . PP 001 SEC.
00.22.58. NASA7A . NASA7A,07,777,100000.
00.22.59. NASA7A . RUN(L)
00.23.00. NASA7A . FORTRAN ERRORS IN    POWER
00.23.00. NASA7A . CP  000.097 SEC.
00.23.00. NASA7A . PP  001.342 SEC.
00.24.29. NASA7A . DUMP.
00.24.32. NASA7A . PP 026 SEC.
00.25.39. NASA7A . READ.
00.25.40. NASA7A . PP 001 SEC.
00.25.41. NASA7A . NASA7A,07,777,100000.
00.25.41. NASA7A . RUN(L)
00.25.42. NASA7A . FORTRAN ERRORS IN    POWER
00.25.42. NASA7A . CP  000.095 SEC.
00.25.42. NASA7A . PP  000.871 SEC.
00.25.43. NASA7A . PRINT.
00.25.54. NASA7A . PP 011 SEC.
OCTOBER TAPE.
```

PROBLEM NO. 7B

```
 *NM                FORTRAN 4 PROGRAM SEVENB

                    PROGRAM SEVENB
 *NM
                    DIMENSION Z(100)
                    DO 2 I=1,100

 000000    7110000001
           10610
           76700
 000001    0400200004
 000002    5110340002    200003
           7120000001
 000003    7130000144
           36612
           37736
 000004    5160340002    200004
           0337360001

                    2 Z(I)=10.

 000005    5140340002    200005
           6214320002
 000006    5150260001
           10650
           56610
 000007    0400200003

                    49 DO 50 I=1,100

 000010    7110000001    200006
           10610
           76700
 000011    0400200011
 000012    5110340002    200010
           7120000001
 000013    7130000144
           36612
           37736
 000014    5160340002    200011
           0337360002

                    J=Z(I)/3.45

 000015    10460
           6214320002
           56510
 000016    5110260002
           45651
           26776
 000017    22677
           5160340003

                    X=FLOAT(2*I-J)-15.3

 000020    7120000002
           10340
           10460
 000021    27702
           27003
           42770
           26607
```

123

```
000022   37664
         27706
         24607
         10560
000023   5110260003
         31651
         24606
000024   5160340004

              Y=X/2.1-FLOAT(MOD(I,J))

         10240
         27003
000025   27602
         24706
         44707
         26777
000026   22777
         27707
         42776
         26607
000027   37636
         5160240002
         10460
000030   27704
         24607
         10260
000031   5150340004
         5110260004
000032   45651
         31662
         24606
000033   5160340005

           50 Z(I)=Z(I)+X*Y+X/Y-20.3

000034   5130340002   200015
         6213320002
000035   56410
         5150340004
000036   5110340005
         10640
         41751
000037   5120260005
         30667
         24606
000040   45751
         30667
         24606
         31662
000041   24606
         56610
         0400200010

              51 GO TO 8

000042   0400360003   200016

              8 STOP

000043   5130260006   200017
         10730
```

124

```
000044  0100400100

          END

000045  5140260006
         10740
000046  0100400200




00.02.10. NASA7B . READ.
00.02.11. NASA7B . PP 000 SEC.
00.02.11. NASA7B . NASA7B,07,777,100000.
00.02.12. NASA7B . RUN(L)
00.02.12. NASA7B . FORTRAN ERRORS IN   *******
00.02.13. NASA7B . CP  000.073 SEC.
00.02.13. NASA7B . PP  000.446 SEC.
00.02.13. NASA7B . PRINT.
00.02.23. NASA7B . PP 009 SEC.
OCTOBER TAPE.
```

Epilogue

No sooner had I finished the NASA project when the Sipros Operating System was "delivered" to CDC and the group was disbanded. Then, I had to return home with my family. We decided to take the "southern road" home: US 10. Our first major stop was the Alamo, where a German was a tour guide and we became good friends as we spoke German. The next major stop we targeted was to stay over in New Orleans. As we entered the access road to the center of town we hit a stretch of dirt road and a passing truck "shot" a good size pebble into my windshield and shattered it. We took the car to the nearest VW dealer and were told that it would take 3 days to have the window fixed. I promptly called my home office and told them what happened. They told me not to rush, but take it easy. In effect we got a 5 day vacation, when the weekend was added to the three days.

Originally we planned to go through Alabama. But, when we hit the road we heard there were some protest marches. So, we decided to take US 10 to US 95 or US 1. First, we visited Oki Fenoki and then we decided to take US 1, the scenic route home. When I called my home office, all hell broke out. Dr. Webb from the NASA evaluation team had called and asked for a "debriefing." He needed a "few questions" answered, and everybody had the jitters. They took his number and told him they would call him back as soon as they located me, since I was in transit.

Then, they called the Alabama and Georgia Police to "please" find me. Of course they never found me after a few hours search. Then, they called Arden Hills, the production center of the 6600 and asked for "Thornton," the production manager and close associate to Seymore Cray. They explained the situation and asked Thornton to call Dr. Webb. He did and answered a few trivial questions (explaining a few instruction).

When we reached New Jersey, a few days later, and I called the office, there was jubilation. My technical response, my solutions and timing analysis had won first prize from NASA! Next day when I arrived at the office, everybody was congratulating me. First, the district manager, then the regional manager, then Norris (the President), and then Seymore Cray. I felt really good when Seymore called. That was the ultimate recognition.

While the office was celebration my success, it was announced that NASA was visiting the Corporate Home Office at 501 Park Avenue. In a jiffy every salesman in my office booked a flight to Minneapolis. Then, a day before the meeting, our office got a call from Dr. Webb. He requested that the "author" of the Technical Proposal be at that meeting. But, when our secretary tried to book a reservation for me all coach seats were taken. Thus, she booked me a first class seat.

When I arrived at the airport I got the "royal treatment" while everybody else waited in line. And, when I got seated I asked the Stewardess for a bottle of Champagne. Then, with the bottle under my arm I walked into the Coach section. I could see how everybody tried to hide behind a paper or a book. But, before I could even mutter a few words, a "Flight Marshall" tapped me on my shoulder and told me:" Sir, you are not permitted in the Coach section. Please return to your seat." And, promptly I obeyed.

Our normal "hangout" in Minneapolis was the Thunderbird Motel. When I arrived an invitation was waiting for me from Dr. Webb. I freshened up and went to see him. He invited we to a Bridge game, which I gladly accepted. I was his partner, while his other 2 associates formed the other pair. It seemed to me that we were playing all night. During the game they were asking me detailed questions about my solutions and timing analysis which only I could answer. But, primarily they were interested to find out if the 6600 was "REAL." My enthusiasm and love for the machine convinced them that it was real and functioning. They told me that I beat most competitors by about a factor of 10 to 1! They say that enthusiasm sells. Well, in that case I sold the 6600 to them. As a consequence more than 100 million Dollar's worth of computers was sold to NASA.

It should be noted that Seymore held that 10 to 1 advantage well beyond his death. The enclosed manual was for the longest time a key selling document for the 6000 series of computers.

127

Later, when I worked for Brookhaven Labs and received the first Software Patent, it prolonged the useful life for the 6000 series by an additional 10 years and more.

Before I took my family to LA, I lived with a packed suitcase in the office. In an emergency I was sent to whatever location needed help. Occasionally I was sent to a secret base. Then, the regional manager would call my wife and tell her: "Your husband is on assignment. I can't tell where he is and he will not be able to call you. But rest assured that he is fine and will return shortly!" After such a call my wife freaked out. This time, when we returned she put her foot down and insisted that I should lead a "normal" life, keeping a normal schedule. When she told that to the Regional Manager and he turned a deaf ear, she began bugging me to leave CDC and find another Job.

It so happened the Brookhaven National Laboratory, in Upton New York was getting a CDC 6600 serial 11. So, I approached the salesman of that account and told him that I was planning to leave CDC. He begged me to take the job at Brookhaven. He would make all the arrangements with the Director of Applied Mathematics, Dr. Y. Shimamoto. When I agreed, he came back with an interview date the next day.

Then, on the appointed day, I arrived at Brookaven and went to see Dr.Shimamoto. His secretary (Amy) led me to his office and asked me to take a seat at his desk. As I got seated I had a clear view of his desk. And there, the only document on his desk was my response to the NASA RFP! Instantly I knew: He was a member of the selection committee who had awarded me the first prize!

Printed in the United States
By Bookmasters